Atlas of
Central Solar Eclipses
in the USA

Fred Espenak

Bifrost Astronomical Observatory

Edition 1.0
January 2016

Atlas of Central Solar Eclipses in the USA

Astropixels Publishing
P.O. Box 16197
Portal, AZ 85632

Astropixels Publishing Website: *astropixels.com/pubs*

This book may be ordered at: *astropixels.com/pubs/SEUSA.html*

Printed in the United States of America

ISBN 978-1-941983-09-6

Astropixels Publication: AP010 (Version 1.0a)

First Edition

Front Cover: A map by Michael Zeiler (GreatAmericanEclipse.com) shows the path of every total solar eclipse passing though the USA from 2001 through 2100. More about central solar eclipses in the USA can be found at:

eclipsewise.com/solar/SEcountry/SE-USA.html

Back Cover Photo of Fred Espenak: Copyright © 2015 by Patricia Espenak,

Table of Contents

SECTION 1: SOLAR ECLIPSE FUNDAMENTALS 5
  1.1 INTRODUCTION 5
  1.2 CLASSIFICATION OF SOLAR ECLIPSES 6
  1.3 CENTRAL SOLAR ECLIPSES 7
  1.4 VISUAL APPEARANCE OF ANNULAR SOLAR ECLIPSES 7
  1.5 VISUAL APPEARANCE OF TOTAL SOLAR ECLIPSES 7

SECTION 2: SOLAR ECLIPSE STATISTICS FOR THE USA 9
  2.1 INTRODUCTION 9
  2.2 TOTAL SOLAR ECLIPSE STATISTICS 9
  2.3 ANNULAR SOLAR ECLIPSE STATISTICS 9
  2.4 HYBRID SOLAR ECLIPSE STATISTICS 10
  2.5 TOTAL SOLAR ECLIPSES BY STATE 10
  2.6 ANNULAR SOLAR ECLIPSES BY STATE 10
  2.7 ECLIPSEWISE AND CENTRAL SOLAR ECLIPSES IN THE USA 10
  2.8 TOTAL SOLAR ECLIPSES IN THE USA FROM 1001 TO 2000 13
  2.9 TOTAL SOLAR ECLIPSES IN THE USA FROM 2001 TO 3000 14

SECTION 3: EXPLANATION OF SOLAR ECLIPSE CATALOG IN APPENDIX A 15
  3.1 INTRODUCTION 15
  3.2 CALENDAR DATE 15
  3.3 TD OF GREATEST ECLIPSE (TERRESTRIAL DYNAMICAL TIME OF GREATEST ECLIPSE) 15
  3.4 ΔT (DELTA T) 15
  3.5 LUNA NUM (LUNATION NUMBER) 16
  3.6 SAROS NUM (SAROS SERIES NUMBER) 16
  3.7 ECL. TYPE (SOLAR ECLIPSE TYPE) 16
  3.7 GAMMA 16
  3.8 ECL. MAG. (ECLIPSE MAGNITUDE) 16
  3.9 LAT. LONG. (LATITUDE AND LONGITUDE) 17
  3.10 SUN ALT (ALTITUDE OF SUN) 17
  3.11 SUN AZM (AZIMUTH OF SUN) 17
  3.12 PATH WIDTH 17
  3.13 CENTRAL LINE DUR. (CENTRAL LINE DURATION) 17
  3.14 USA GEOGRAPHIC REGION 17

SECTION 4: EXPLANATION OF GLOBAL SOLAR ECLIPSE MAPS IN APPENDIX B 18
  4.1 INTRODUCTION 18
  4.2 SOLAR ECLIPSE TYPE 19
  4.3 SAROS SERIES NUMBER 19
  4.4 NODE 19
  4.5 CALENDAR DATE 20
  4.6 GREATEST ECLIPSE 20
  4.7 ΔT (DELTA T) 20
  4.8 GAMMA 20
  4.9 ALTITUDE OF SUN 20
  4.10 DURATION OF CENTRAL ECLIPSE 20
  4.11 ECLIPSE MAGNITUDE 20

SECTION 5: EXPLANATION OF MAPS IN LOWER 48 STATES IN APPENDIX C...................................**21**
  5.1 INTRODUCTION 21

SECTION 6: SOLAR ECLIPSE PREDICTIONS...**23**
  6.1 MEAN LUNAR RADIUS 23
  6.2 SOLAR AND LUNAR COORDINATES 23
  6.3 MEASUREMENT OF TIME 23
  6.4 ΔT (DELTA T) 24
  6.5 CALENDAR DATE 24

APPENDIX A..**25**
  CATALOG OF CENTRAL SOLAR ECLIPSES IN THE USA: 1001 – 3000 25
  KEY TO SOLAR ECLIPSE CATALOG 26

APPENDIX B..**37**
  GLOBAL MAPS OF CENTRAL SOLAR ECLIPSES IN THE USA 37
  KEY TO SOLAR ECLIPSE MAPS 38

APPENDIX C..**81**
  MAPS OF CENTRAL SOLAR ECLIPSES IN THE LOWER 48 STATES OF THE USA 81

BIBLIOGRAPHY..**103**

ASTROPIXELS PUBLICATIONS..**104**

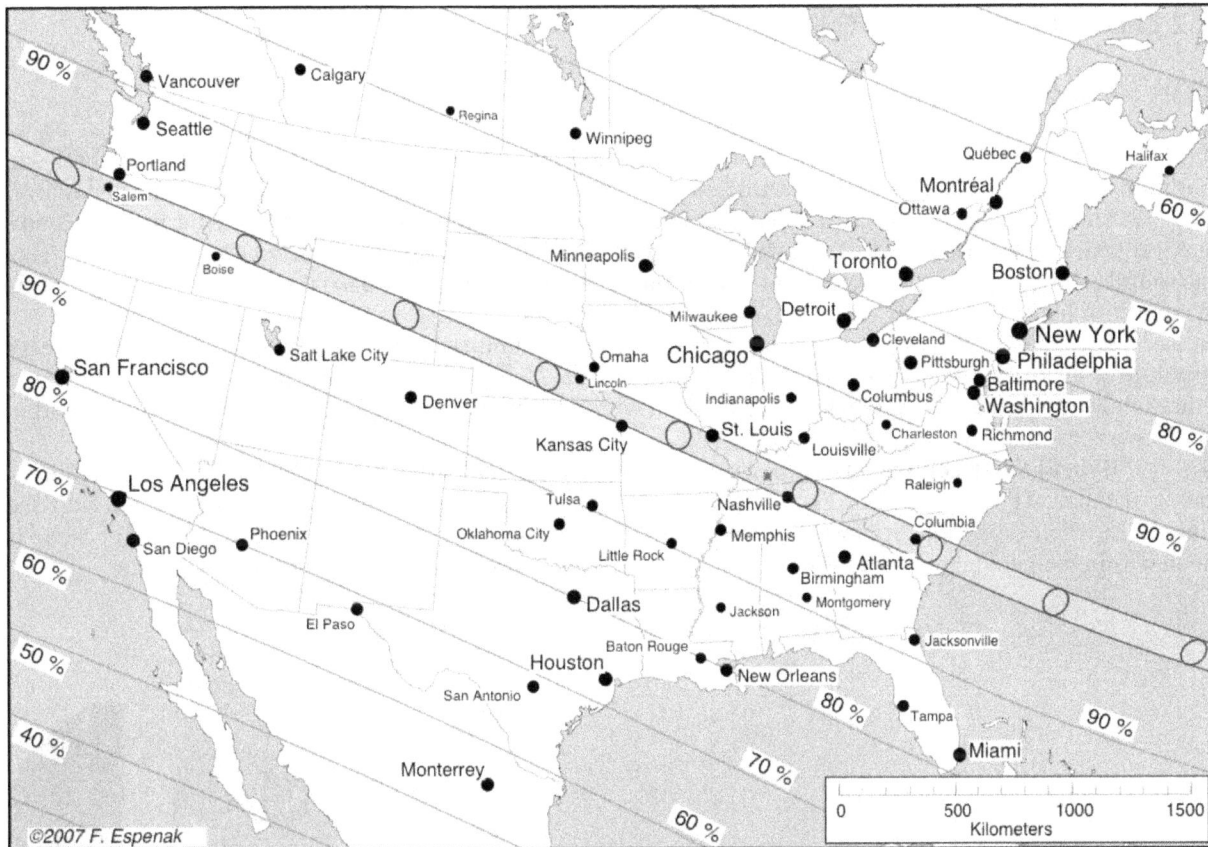

Map 1–1 shows the path of the total solar eclipse of August 21, 2017 though the USA.
The curves marked by percentages indicate the eclipse magnitude at other locations in the USA.

Section 1: Solar Eclipse Fundamentals

1.1 Introduction

A total eclipse of the Sun is arguably the most spectacular astronomical event visible to the naked eye. For a few brief minutes, the Sun's brilliant disk is hidden from view and the faint solar corona surrounding the Sun is revealed. The startling event is accompanied by an eerie twilight in which brighter stars and planets are visible in the daytime sky, and the colors of sunset surround the entire horizon.

Although a total solar eclipse occurs once every 12 to 18 months, it is only visible from within the narrow track of the Moon's umbral shadow as it races across Earth's surface. On average, this track or path of totality is only 140 miles (225 km) wide encompassing but a tiny fraction of Earth's surface. It is for this reason that total eclipses seem to be so rare from any one location.

With interest rapidly building for the upcoming total solar eclipse visible from the USA on August 21, 2017[1] (Map 1–1), it is quite natural that questions will arise about the rarity of the eclipse — when was the last one and when is the next. That is the subject of this book.

---

[1] for complete details see "Eclipse Bulletin: Total Solar Eclipse of 2017 August 21", astropixels.com/pubs/TSE2017.html

1.2 Classification of Solar Eclipses

There are four basic types of solar eclipses:

1. **Partial Solar Eclipse** – The Moon's penumbral shadow traverses Earth. The Moon's umbral and antumbral shadows completely miss Earth. A portion of the Sun's disk is obscured from within the penumbra.
2. **Annular Solar Eclipse** – The Moon's penumbral and antumbral shadows traverse Earth. The Moon's umbral shadow completely misses Earth. The Moon's disk appears smaller than the Sun so a bright ring surrounds the Moon when viewed from within the antumbral shadow. A partial eclipse is seen within the penumbral shadow.
3. **Total Solar Eclipse** – The Moon's penumbral and umbral shadows traverse Earth. The Moon's antumbral shadow extends beyond Earth's surface. The Moon's disk appears larger than the Sun and completely covers the solar disk when viewed from within the umbral shadow. A partial eclipse is seen within the penumbral shadow.
4. **Hybrid Solar Eclipse** – The Moon's penumbral, umbral and antumbral shadows all traverse different parts of Earth. The curvature of Earth's surface brings some regions into the umbra and others into the antumbra. The eclipse appears total within the umbra and annular within the antumbra. A partial eclipse is seen within the penumbral shadow. Hybrid eclipses are also known as annular-total eclipses.

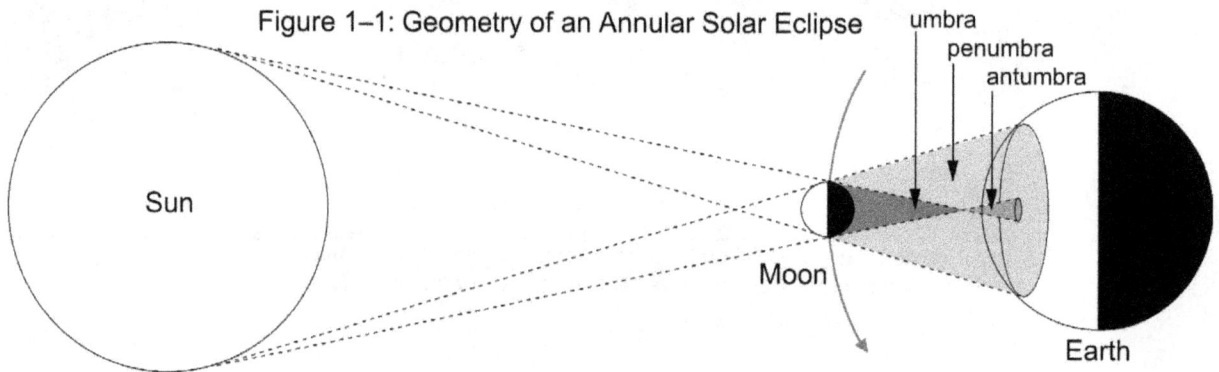

Figure 1–1: Geometry of an Annular Solar Eclipse

Figure 1–1 illustrates the geometry of an annular solar eclipse. A partial eclipse is visible from within the large penumbral shadow, while the annular eclipse is confined to the much smaller antumbral shadow.

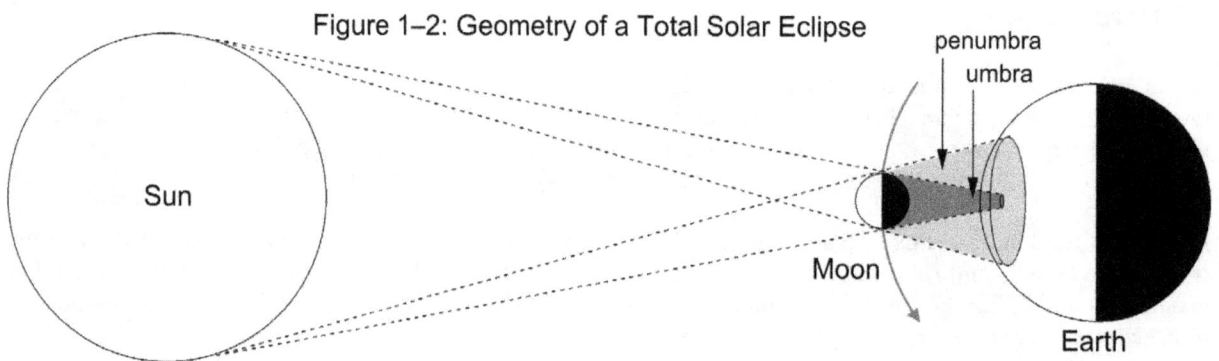

Figure 1–2: Geometry of a Total Solar Eclipse

Figure 1–2 illustrates the geometry of a total solar eclipse. A partial eclipse is visible from within the large penumbral shadow, while the total eclipse is only seen from the much smaller umbral shadow.

1.3 Central Solar Eclipses

When the central axis of the Moon's umbral or antumbral shadow strikes Earth, the resulting eclipse is either a total, annular or hybrid. Thus, the term central[2] eclipse is a convenient way of grouping these three eclipses into a single category. In contrast, the central shadow axis always misses Earth during a partial eclipse.

The Moon casts a second shadow called the penumbra that is approximately 6700 to 7300 kilometers (4200 to 4500 miles) in diameter and can cover a significant fraction of the daytime hemisphere of Earth. Anyone inside the penumbra will observe a partial solar eclipse. Consequently, all central eclipses begin and end with a partial eclipse visible from large geographic areas as the penumbra sweeps across Earth's surface.

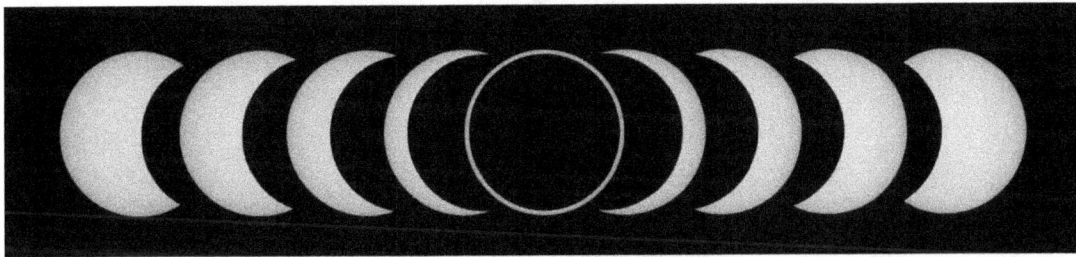

Photo 1–1 shows various phases of the Annular solar eclipse of 2005 Oct 03. ©2005 F. Espenak

1.4 Visual Appearance of Annular Solar Eclipses

During an annular eclipse, the Moon's penumbral and antumbral shadows sweep across Earth. While the penumbra is quite large, the antumbra is much smaller and has a maximum diameter of 374 kilometers (232 miles). Because of this, the antumbra covers a much smaller fraction of Earth's surface.

The Moon's orbital motion carries the penumbral and antumbral shadows in a west to east direction. A partial eclipse is visible within the penumbra, but only observers located in the much narrower path of the antumbra will see an annular eclipse. The antumbra's track across Earth is called the path of annularity.

All annular eclipses begin with a series of partial phases lasting up to an hour or more. At the peak of the eclipse, the Moon's disk can be seen in complete silhouette against the Sun. The remaining solar photosphere appears as an intensely bright ring of light surrounding the Moon. The annular phase lasts a maximum of 12 ½ minutes, but it is more typically 3 to 6 minutes in length. After annularity, another series of partial phases occur as the Moon gradually uncovers the Sun.

Even during the annular phase, the Sun is dangerously bright and cannot be viewed without a solar filter. In this regard, annular eclipses are similar to partial eclipses. The landscape and sky remain bright throughout the eclipse, giving little indication of the celestial event in progress.

1.5 Visual Appearance of Total Solar Eclipses

During a total eclipse, the Moon's penumbral and umbral shadows fall upon Earth. The umbra is much smaller than the penumbra and has a maximum diameter of 273 kilometers (170 miles). The narrow track traced out by the umbra as it sweeps west to east across Earth's surface is called the path of totality. Anyone standing within this zone will see the Sun completely obscured by the Moon. The total phase can last up to 7 ½ minutes, but is more typically 2 to 3 minutes in length.

---

[2] Under rare circumstances, it is possible to have a total or annular eclipse that in not central. Such events occur at high latitudes when the umbral or antumbral shadow just grazes Earth while the central axis misses. Such events are classified as non-central total (0.8%) or non-central annular eclipses (1.8%).

Total eclipses all begin and end with a series of partial phases lasting up to an hour or more. But this is where the resemblance between partial and annular eclipses ends. The total phase is the most spectacular astronomical event visible to the naked eye. At this time, the Sun's outer atmosphere — the solar corona — appears as a gossamer halo surrounding the Moon, and bright stars and planets are visible.

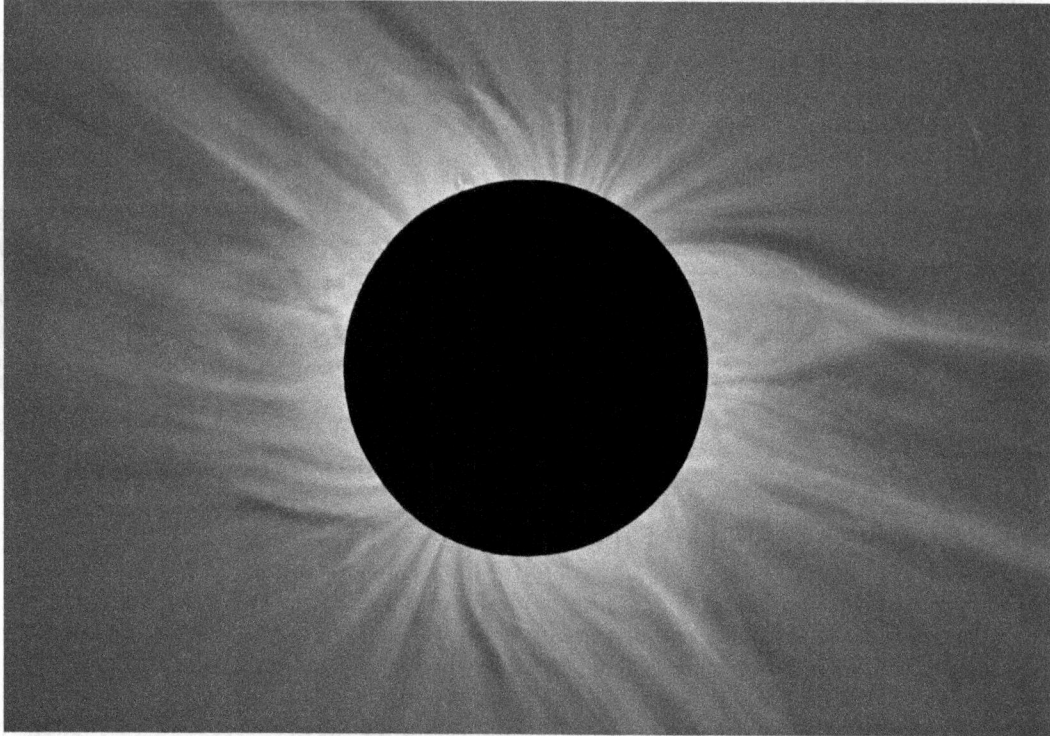

Photo 1–2 captures details in the solar corona during the total eclipse of 2006 Mar 29. ©2006 F. Espenak

The eclipse starts to take on a unique character about five minutes before the total phase begins. Sunlight has a foreboding quality and casts abnormally sharp shadows. The approaching lunar umbra darkens the western sky and the air temperature is noticeably cooler. A minute before totality, ghostly shadow bands[3] ripple across the ground. The ambient light grows feeble even though the crescent Sun is still too bright to look at. In the final seconds, the Sun's corona emerges from the glare as the solar crescent shrinks to a single brilliant jewel. This celestial diamond ring lingers for a moment before the last bead of sunlight is extinguished.

The Sun's glorious corona is now displayed to full advantage in the darkened sky of totality. The simple act of standing within the shadow of the Moon affords the rare opportunity to gaze directly at the glowing million-degree plasma surrounding our star. Twisted and constrained by the Sun's enormous magnetic fields, the solar corona is revealed to the naked eye only during the brief seconds when the Moon completely blocks the Sun's brilliant disk. An eerie twilight bathes the landscape and the colors of dusk surround the horizon.

The minutes race by like seconds. Suddenly, a sparkling bead of sunlight reappears along one edge of the Moon and quickly grows to blindingly bright proportions. Daylight returns as the corona fades and the total phase ends. Although another hour of partial phases remains before the eclipse ends, these are anticlimactic after the spectacle of totality.

While filters are required for viewing the partial phases, they must be removed for totality. The total phase is the only time it is completely safe to look directly at the Sun without protection. In fact, the total phase is not even visible through solar filters because the Sun's corona is a million times fainter than the photosphere.

---

[3] Shadow bands are often seen immediately before and after a total solar eclipse. They appear as thin parallel wavy lines of alternating light and dark that race across the ground.

Section 2: Solar Eclipse Statistics for the USA

2.1 Introduction

During the 2000-year period from 1001 to 3000, there are 4773 solar eclipses. Of these, 1684 are partial while the remaining 3089 are either total, annular, or hybrid (aka *central*[4]). They may be broken down as follows:

Table 2-1: Statistics for All Central Eclipses: 1001 to 3000

Type	Number	% of All
T	1266	41.0%
A	1601	51.8%
H	200	6.5%

Among these 3089 central eclipses, 499 (16.2%) cross through some portion of the USA. For the purposes of this *Atlas*, the USA has been divided into three separate geographic regions:

1) Lower 48 States (L48) – The contiguous United States consisting of the 48 adjoining U.S. states and Washington, DC
2) Alaska (AK) – the state of Alaska (including the Aleutian Islands)
3) Hawaii (HI) – the state of Hawaii (limited to the eight main islands)

None of the sixteen territories of the United States are considered in this *Atlas*.

2.2 Total Solar Eclipse Statistics

Of the 499 central solar eclipses, 206 (41.3%) of them are total. They may be broken down into the three geographic regions as follows:

Table 2-2: Statistics for Total Eclipses

Region	Number[†]
L48	150
AK	82/31
HI	8/3

[†] when two numbers are listed, the first is the total number for that region,
while the second number is the number of eclipses also visible from L48

2.3 Annular Solar Eclipse Statistics

Of the 499 central solar eclipses, 264 (52.9%) of them are annular. They may be broken down into the three geographic regions as follows:

Table 2-3: Statistics for Annular Eclipses

Region	Number[†]
L48	183
AK	91/32
HI	29/7

[†] when two numbers are listed, the first is the total number for that region,
while the second number is the number of eclipses also visible from L48

---

[4] A small fraction of total and annular eclipses are actually non-central. Such events occur at high latitudes when the umbral or antumbral shadow just grazes Earth while the central axis misses. For the purposes of this publication, we will ignore this distinction and classify all total, annular and hybrid eclipses as central.

2.4 Hybrid Solar Eclipse Statistics

Of the 499 central solar eclipses, 29 (5.8%) of them are hybrid. They may be broken down into the three geographic regions as follows:

Table 2–4: Statistics for Hybrid Eclipses

Region	Number[†]
L48	21
AK	9/3
HI	2/0

[†] when two numbers are listed, the first is the total number for that region, while the second number is the number of eclipses also visible from L48

2.5 Total Solar Eclipses By State

Table 2–5 lists the year of total solar eclipses passing through each of the 50 states in the USA as well as Washington, DC. The eclipses are arranged in two groups: (1601–2000) and (2001–2300) total eclipses. Years in square brackets "[ ]" fall outside the nominal year range of the table. These tables can be used to quickly determine when the last or next total eclipse occurs from any state.

2.6 Annular Solar Eclipses By State

Table 2–6 lists the year of annular solar eclipses passing through each of the 50 states in the USA as well as Washington, DC. The eclipses are arranged in two groups: (1801–2000) and (2001–2150) annular eclipses. Years in square brackets "[ ]" fall outside the nominal year range of the table. These tables can be used to quickly determine when the last or next annular eclipse occurs from any state.

The date range of these two tables is a small subset of the 1000 to 3000 time period in order to fit on a single page. There may be other eclipses visible from each state not listed but falling outside the nominal date range of each table.

2.7 EclipseWise and Central Solar Eclipses in the USA

EclipseWise.com has a complementary web page for central solar eclipses in the USA. It offers links to larger versions of the global maps in Appendix B as well a zoomable Google map for every eclipse in the *Atlas*.

eclipsewise.com/solar/SEcountry/SEinUSA.html

Another useful feature is the JavaScript Solar Eclipse Explorer. This is an interactive web page that can calculate the local circumstances for all solar eclipses visible from any city or geographic location.

eclipsewise.com/solar/JSEX/JSEX-USA.html

Note that the mean frequency of a total solar eclipse from any given place is once in 375 years (Meeus, 1982). Because the average width of an annular eclipse track is broader and covers a larger area, the mean frequency of an annular eclipse from any given place is shorter at once in 224 years.

Table 2–5: Total Eclipses By State

US State	Total Eclipses: 1601-2000	Total Eclipses: 2001-2300
Alabama	1623,1778,1834,1900,1918	2045,2052,2078,2200
Alaska	1632,1672,1742,1869,1878,1889,1963	2108,2153,2169
Arizona	1623,1679,1724,1806	2205,2207,2245
Arkansas	1623,1834,1918	2024,2045
California	1623,1632,1677,1679,1724,1880,1889,1923	2045,2106,2205,2207,2252,2254
Colorado	1618,1679,1724,1742,1806,1834,1878,1918	2045,2106,2169,2205,2245
Connecticut	1806,1925	2079,2144,2200
Delaware	[1478]	2144
Florida	1625,1713,1752,1778,1803,1825,1918,1970	2045,2052,2078,2198,2259,2261
Georgia	1623,1778,1834,1900,1970	2017,2045,2052,2078,2153,2198,2298
Hawaii	1670,1679,1991	2252,2254
Idaho	1618,1742,1878,1889,1918,1979	2017,2169,2252,2254
Illinois	1717,1806,1869	2017,2024,2153,2178,2205
Indiana	1717,1806,1869	2024,2099,2153
Iowa	1717,1806,1869,1954	2153,2178,2205,2245
Kansas	1618,1806,1834,1878,1918	2017,2045,2169,2205,2245
Kentucky	1717,1869	2017,2024,2153,2200
Louisiana	1618,1623,1768,1778,1900,1918	2045,2052,2078,2198,2200
Maine	1659,1672,1780,1932,1963	2024,2079,2106,2200
Maryland	1778	2144,2200
Mass.	1684,1806,1932,1959,1970	2200
Michigan	1558,1806,1925	2099,2144,2205,2263
Minnesota	1679,1724,1869,1925,1954	2099,2144,2153,2245,2263
Mississippi	1623,1768,1778,1834,1900,1918	2045,2200
Missouri	1717,1806,1869	2017,2024,2178,2205
Montana	1717,1834,1869,1878,1889,1945,1979	2044,2099,2153,2178,2252,2254
Nebraska	1679,1834,1869,1954	2017,2106,2178,2205,2245
Nevada	1648,1677,1724,1742,1880,1889	2045,2106,2169,2207,2252,2254
N. Hampshire	1672,1806,1932,1959	2024,2079,2200
New Jersey	1925	2079,2144,2200
New Mexico	1623,1742,1878	2169,2205,2207,2245
New York	1806,1925	2024,2079,2144,2200,2205,2263
N. Carolina	1623,1717,1778,1869,1900,1970	2017,2078,2099,2153
N. Dakota	1679,1717,1724,1869,1889,1979,2044,2099	2044,2099,2178,2252,2254
Ohio	1806	2024,2099,2144
Oklahoma	1618,1806,1834,1878,1918	2024,2045,2169,2205,2245
Oregon	1618,1648,1742,1918,1979	2017,2169,2207,2254
Pennsylvania	1806,1925	2024,2079,2099,2144,2200,2263
Rhode Island	1806,1925	2079
S. Carolina	1623,1717,1778,1834,1869,1900,1970	2017,2052,2078,2153
S. Dakota	1679,1717,1724,1869,1954	2106,2153,2178
Tennessee	1717,1869	2017,2153,2200
Texas	1618,1623,1713,1768,1778,1806,1878,1900	2024,2045,2052,2078,2169,2198,2200
Utah	1677,1679,1724,1742,1880,1918	2045,2106,2169
Vermont	1806,1932	2024,2079
Virginia	1778,1869,1900,1970	2078,2099,2153
Washington	1618,1670,1860,1878,1918,1979	2169
Washington, DC	[1451,1478]	[2444]
West Virginia	1869	2099,2200
Wisconsin	1925	2099,2106,2144,2205,2245,2263
Wyoming	1618,1679,1724,1834,1878,1880,1889,1918	2017,2106,2169,2252,2254

Table 2-6: Annular Eclipses By State

US State	Annular Eclipses: 1801-2000	Annular Eclipses: 2001-2150
Alabama	1821,1865,1940,1984	[2238,2267]
Alaska	1809,1811,1849,1856,1939,1948	2039,2084,2122,2137
Arizona	1821,1994	2012,2023,2077,2121
Arkansas	1821	[2238,2267]
California	1822,1885,1992	2012,2023,2046,2077,2100,2110,2121,2131
Colorado	1865	2012,2023,2048,2077,2121
Connecticut	1854,1875	[2267]
Delaware	1811,1831,1838	2111 [2267]
Florida	1908,1940	[2238,2240]
Georgia	1821,1831,1865,1940,1984	[2238,2267]
Hawaii	1829,1839,1876	2046,2149
Idaho	1822,1854,1865,1885	2046,2077,2100,2110,2111
Illinois	1811,1865,1994	2048,2093,2111,2121
Indiana	1811,1838,1865,1994	2093,2111,2121
Iowa	1865	2048
Kansas	1865	2048,2121
Kentucky	1865	2121
Louisiana	1821,1831,1940,1984	[2238]
Maine	1854,1875,1994	2093
Maryland	1811,1831,1838	2111
Mass.	1831,1854,1875,1994	[2247,2267]
Michigan	1811,1838,1854,1994	2048,2057,2093,2111
Minnesota	1811,1838,1854	2048,2100,2110,2111
Mississippi	1831,1940,1984	[2238,2267]
Missouri	1865,1994	2048,2121
Montana	1822,1854,1865,1885	2100,2110,2111
Nebraska	1865	2048
Nevada	1822	2012,2023,2046,2077,2100,2110
N. Hampshire	1854,1875,1994	2093
New Jersey	1831	[2267]
New Mexico	1821,1994	2012,2023,2077,2121
New York	1838,1854,1875,1994	2093
N. Carolina	1811,1831,1838,1865,1951,1984	[2267]
N. Dakota	1854	2100,2110,2111
Ohio	1811,1838,1994	2093,2111
Oklahoma	1994	[2251,2267]
Oregon	1822,1865,1885	2012,2023,2046,2077,2100,2110
Pennsylvania	1811,1838,1994	2093,2111
Rhode Island	[1791]	[2267]
S. Carolina	1821,1831,1865,1984	[2238,2267]
S. Dakota	1865	2100,2110,2111
Tennessee	1865	[2238,2251,2267,2294]
Texas	1821,1831,1919,1940,1984,1994	2012,2023,2056,2077
Utah	[1782]	2012,2023,2077
Vermont	1854,1875,1994	2093
Virginia	1811,1831,1838,1951,1984	2111
Washington	1854,1865	2077,2111
Washington, DC	1854,1865	2077,2111
West Virginia	1811,1838	2111,2121
Wisconsin	1811,1838	2048,2111
Wyoming	1865	2100,2110

2.8 Total Solar Eclipses in the USA from 1001 to 2000

During the Second Millennium (1001 to 2000) was there any location in the USA's lower 48 states that was not in the path of a total solar eclipse? It's a simple question, but it requires computing and plotting every total eclipse during this period on a map as shown in Figure 2–1.

The vast majority of this region as well as Canada and Mexico fall within the path of at least one total eclipse. Some notable exceptions include the following.

1) eastern border of Maine and southern New Brunswick
2) southeastern Illinois and southwestern Ohio (including Dayton)
3) southeastern Ohio
4) southwestern Georgia
5) several long tracks in Oklahoma and Arkansas
6) central western Louisiana
7) southwestern Texas (including El Paso)
8) the "bootheel" of New Mexico
9) southeastern Arizona (including the author's home in Portal)
10) a large portion of central Nebraska
11) two regions in northern Colorado
12) southeastern corner of Wyoming (including Cheyenne)
13) several regions in Saskatchewan and Quebec

There is no particular reason why these areas have been "shunned" by the Moon's umbral shadow. They have all been in the path of total eclipses prior to 1001 CE. In fact, many of them will fall within the path of the umbral shadow in the next millennium.

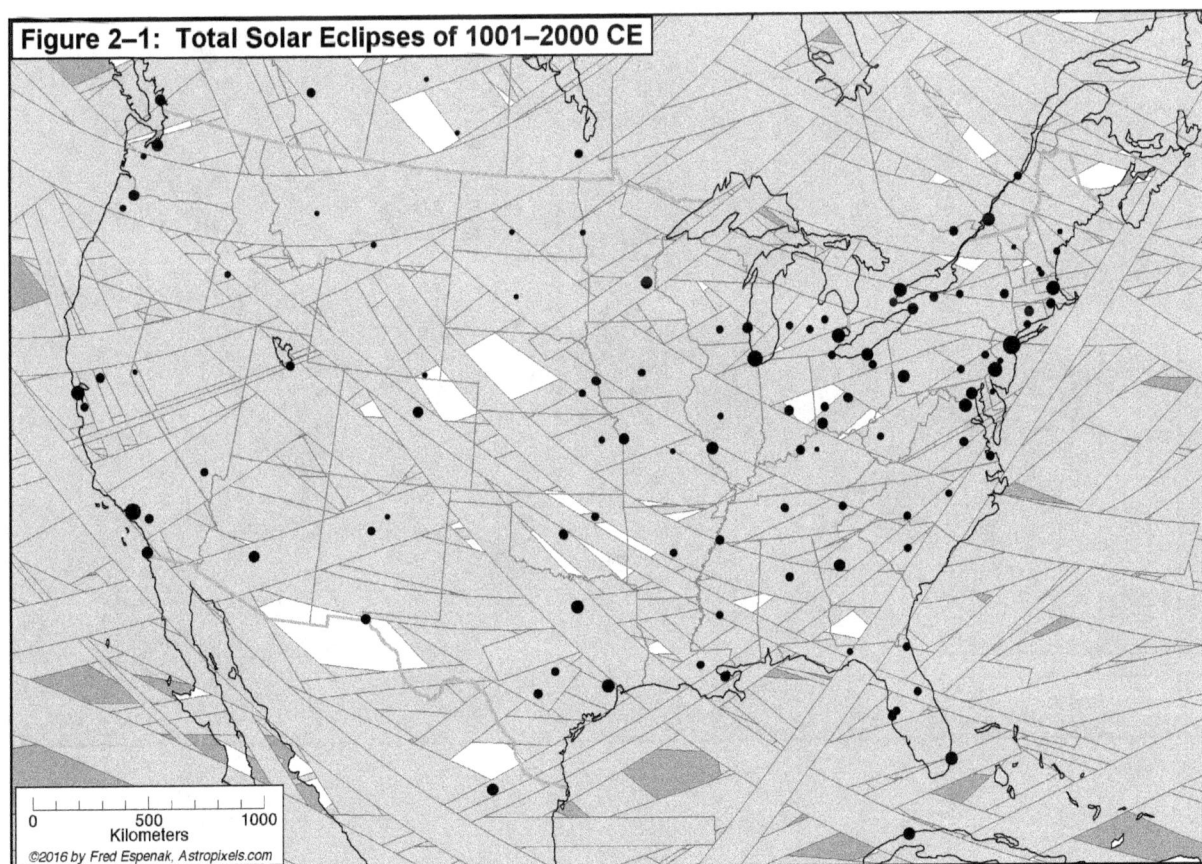

Figure 2–1: Total Solar Eclipses of 1001–2000 CE

©2016 by Fred Espenak, Astropixels.com

2.9 Total Solar Eclipses in the USA from 2001 to 3000

During the Third Millennium (2001 to 3000) is there any location in the USA's lower 48 states that will not be in the path of a total solar eclipse? Once again the answer requires computing and plotting every total eclipse during this period on a map as shown in Figure 2–2.

Most of the lower 48 states, Canada and Mexico fall within the path of at least one total eclipse. Some notable exceptions include the following.

1) northern Virginia
2) southern Ohio, northern Kentucky and western West Virginia (including Cincinnati)
3) two regions in central Georgia
4) southern Arkansas and northern Louisiana
5) three large regions in Texas (including Houston)
6) two regions in New Mexico
7) eastern Arizona
8) southwestern Nebraska and northeastern Colorado
9) southwestern Wyoming and southeastern Idaho
10) western Montana

Since the mean frequency of a total solar eclipse from any given place is once every 375 years (Meeus, 1982), all locations will eventually fall within the path of a total eclipse given a large enough time interval.

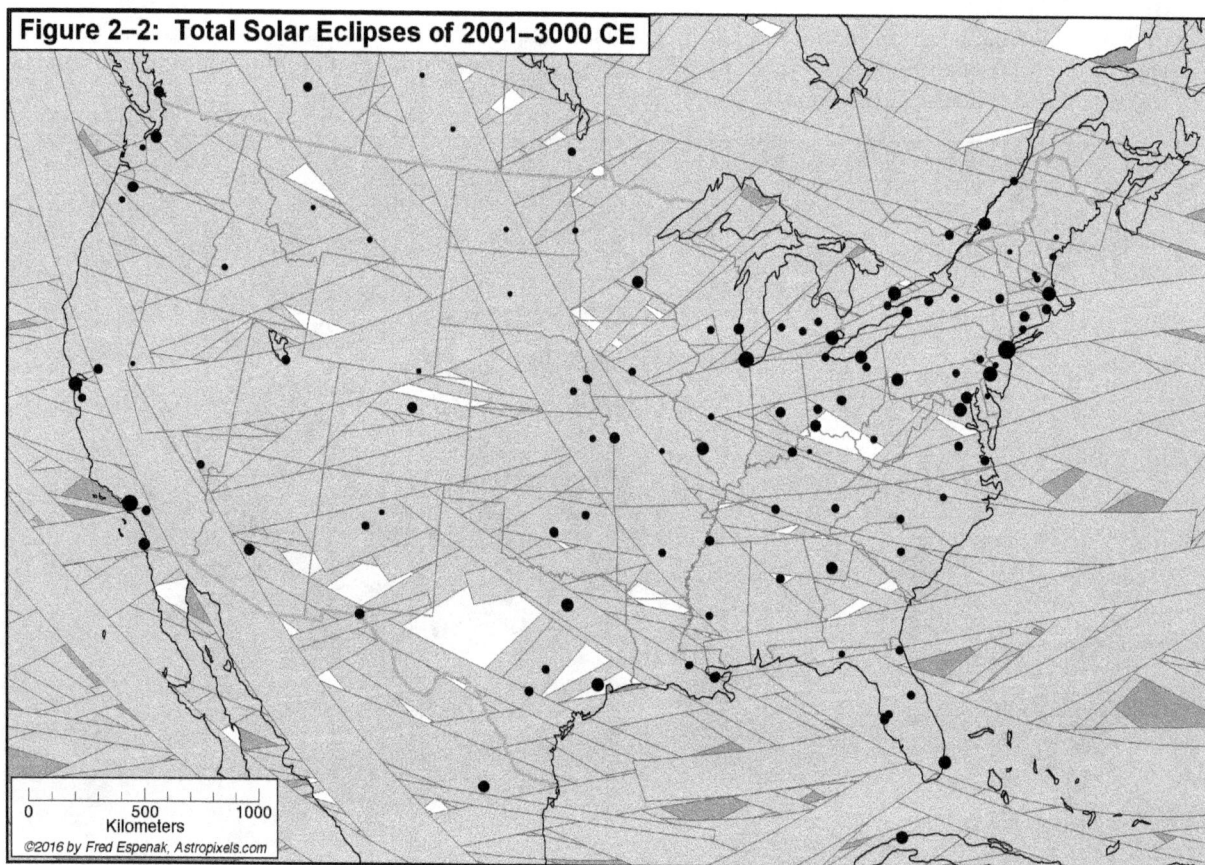

Figure 2–2: Total Solar Eclipses of 2001–3000 CE

Section 3: Explanation of Solar Eclipse Catalog in Appendix A

3.1 Introduction

The USA experiences 499 central eclipses of the Sun during the 2000-year period from 1001 to 3000. The catalog in Appendix A consists of a series of tables that summarize the principal characteristics of each solar eclipse over this time interval. The tables compliment the eclipse maps in Appendix B.

For the purposes of this Atlas, the USA consists of three separate geographic regions
1) Lower 48 States (L48) – The contiguous United States consisting of the 48 adjoining U.S. states and Washington, DC
2) Alaska (AK) – the state of Alaska (including the Aleutian Islands)
3) Hawaii (HI) – the state of Hawaii (limited to the eight main islands)
4) None of the sixteen territories of the United States are included in this *Atlas*

Each line in the catalog corresponds to one eclipse and provides concise parameters to characterize the eclipse. The calendar date and Dynamical Time of the instant of greatest eclipse are given along with the adopted value of delta T (ΔT). The lunation number and the Saros series are listed along with the eclipse type. Gamma is the distance of the shadow axis from Earth's center at greatest eclipse, while the eclipse magnitude is defined as the fraction of the Sun's diameter obscured at that instant. The geographic latitude and longitude of the umbra are given for greatest eclipse, along with the Sun's altitude and azimuth, the width of the path, and the central line duration of totality or annularity. The last field indicates which geographic regions of the USA lie within the central eclipse path: L48, AK and/or HI. A more detailed description of each field in the catalog appears in the following sections.

3.2 Calendar Date

The Julian calendar is used prior to 1582 Oct 15. All eclipse dates from 1582 Oct 15 onwards use the modern Gregorian calendar currently found throughout most of the world. Because of the Gregorian Calendar Reform, the day following 1582 Oct 04 (Julian calendar) is 1582 Oct 15 (Gregorian calendar), see Section 6.5.

3.3 TD of Greatest Eclipse (Terrestrial Dynamical Time of Greatest Eclipse)

The instant of greatest eclipse occurs when the distance between the axis of the Moon's shadow cone and the center of Earth reaches a minimum. For partial eclipses, the instant of greatest eclipse differs slightly from the instant of greatest magnitude due to Earth's flattening. For total eclipses, the instant of greatest eclipse differs slightly from the instant of greatest duration, although the differences are quite small.

Greatest eclipse is given in Terrestrial Dynamical Time or TD (Section 6.3), which is a time system based on International Atomic Time. As such, TD is the modern equivalent to its predecessor Ephemeris Time and is used in the theories of planetary motion in the Solar System. To determine the geographic visibility of an eclipse, TD is converted to Universal Time (UT1) using the parameter ΔT (Section 6.4).

3.4 ΔT (Delta T)

Delta T (ΔT) is the arithmetic difference, in seconds, between Terrestrial Dynamical Time (TD) and Universal Time (UT1). For more information on ΔT, see Section 6.4.

3.5 Luna Num (Lunation Number)

The number of lunations (synodic months) elapsed since 2000 Jan 06.

3.6 Saros Num (Saros Series Number)

Each eclipse belongs to a Saros series using a numbering system first introduced by van den Bergh (1955). The eclipses with an odd Saros number take place at the ascending node of the Moon's orbit; those with an even Saros number take place at the descending node.

The Saros is a period of 223 synodic months, or approximately 18 years, 11 days, and 8 hours. Eclipses separated by this period belong to the same Saros series and share similar geometry and characteristics.

3.7 Ecl. Type (Solar Eclipse Type)

The first character in this 2-character parameter gives the eclipse type. The four basic types of solar eclipses are:

1. A = Annular Solar Eclipse – Moon's antumbral shadow traverses Earth; Moon is too far from Earth to completely cover the Sun
2. T = Total Solar Eclipse – Moon's umbral shadow traverses Earth; Moon is close enough to Earth to completely cover the Sun
3. H = Hybrid Solar Eclipse – Moon's umbral and antumbral shadows traverse different parts of Earth; eclipse appears either total or annular along different sections of its path

The second character of the eclipse type is a qualifier defined as follows.

1. n = Central eclipse with no northern limit
2. s = Central eclipse with no southern limit
3. + = Non-central eclipse with no northern limit
4. – = Non-central eclipse with no southern limit

3.7 Gamma

Gamma is the minimum distance from the axis of the lunar shadow cone to the center of Earth, in units of Earth's equatorial radius. This distance is positive or negative, depending on whether the axis of the shadow cone passes north or south of Earth's center. If gamma is between +0.997 and –0.997, the eclipse is central (either total, annular, or hybrid). The limiting value 0.997 differs from unity because of the flattening of Earth.

3.8 Ecl. Mag. (Eclipse Magnitude)

The eclipse magnitude is defined as the fraction of the Sun's diameter occulted by the Moon. For partial eclipses, the eclipse magnitude at the instant of greatest eclipse is given for the geographic position closest to the axis of the Moon's shadow cone. For central eclipses (total, annular, and hybrid), the eclipse magnitude listed is actually the ratio of the topocentric apparent diameters of the Moon and Sun at greatest eclipse. The eclipse magnitude is always less than 1.0 for partial and annular eclipses, but equal to, or greater than, 1.0 for total and hybrid eclipses.

3.9 Lat. Long. (Latitude and Longitude)

The latitude and longitude corresponds to the geographic position of greatest eclipse.

3.10 Sun Alt (Altitude of Sun)

The Sun's altitude at the geographic position intersected by the axis of the lunar shadow cone is given at the instant of greatest eclipse. For partial eclipses, the Sun's altitude is always 0° because the shadow axis misses Earth. In this case, the geographic position corresponds to the point closest to the shadow axis.

3.11 Sun Azm (Azimuth of Sun)

The Sun's azimuth at the geographic position intersected by the axis of the lunar shadow cone is given at the instant of greatest eclipse. The values 0°, 90°, 180°, and 270° correspond to the cardinal directions north, east, south and west, respectively.

3.12 Path Width

For central eclipses (total, annular, or hybrid), the width of the path of totality or annularity (kilometers) is given at the geographic position intersected by the axis of the lunar shadow cone at the instant of greatest eclipse.

3.13 Central Line Dur. (Central Line Duration)

The central line duration of the total or annular phase (in minutes and seconds) is given at the geographic position intersected by the axis of the lunar shadow cone at the instant of greatest eclipse.

In the case of total and hybrid eclipses, this duration is the maximum duration (ignoring limb profile effects) of the total phase along the entire umbral path. For annular eclipses, the duration at greatest eclipse may be near either the minimum or maximum duration of the annular phase along the path. If the annular phase duration exceeds approximately 2.3 min, then it corresponds to the maximum duration along the central line track. If the annular phase duration is less, however, then it corresponds to a minimum and the annular duration increases towards the ends of the central path.

3.14 USA Geographic Region

1) Lower 48 States (L48) – The contiguous United States consisting of the 48 adjoining U.S. states and Washington, DC
2) Alaska (AK) – the state of Alaska (including the Aleutian Islands)
3) Hawaii (HI) – the state of Hawaii (limited to the eight main islands)

Section 4: Explanation of Global Solar Eclipse Maps in Appendix B

4.1 Introduction

The USA experiences 499 central eclipses of the Sun during the 2000-year period from 1001 to 3000. An individual global map for each eclipse appears in Appendix B.

The geographic visibility of each eclipse is illustrated with an orthographic projection map of Earth showing the path of the Moon's penumbral (partial) and umbral/antumbral (total, hybrid, or annular) shadows with respect to the continental coastlines, political boundaries (circa 2000) and the Equator. North is to the top and the daylight terminator is drawn for the instant of greatest eclipse. An x symbol marks the sub-solar point where the Sun appears directly overhead at that time. The salient features of the eclipse maps are identified in Figure 5–1, which serves as a key.

Figure 4–1: Key to Global Solar Eclipse Maps

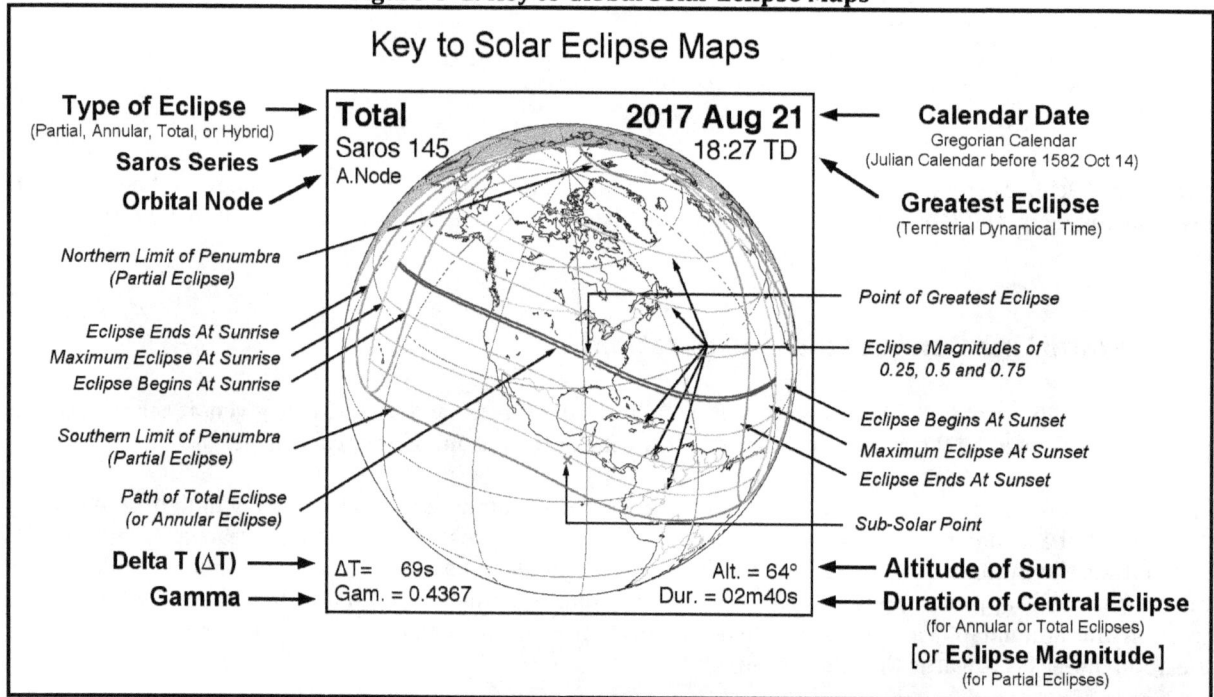

The limits of the Moon's penumbral shadow delineate the region of visibility of a partial solar eclipse. This irregular or saddle shaped region often covers more than half the daylight hemisphere of Earth and consists of several distinct zones or limits. At the northern and/or southern boundaries lie the limits of the penumbra's path. Partial eclipses have only one of these limits, as do central eclipses when the Moon's shadow axis falls no closer than about 0.45 radii from Earth's center. Great loops at the western and eastern extremes of the penumbra's path identify the areas where the eclipse begins/ends at sunrise and sunset, respectively. If the penumbra has both a northern and southern limit, the rising and setting curves form two separate, closed loops. Otherwise, the curves are connected in a distorted figure eight. Bisecting the "eclipse begins/ends at sunrise and sunset" loops is the curve of maximum eclipse at sunrise (western loop) and sunset (eastern loop).

The eclipse magnitude is defined as the fraction of the Sun's diameter occulted by the Moon. The curves of eclipse magnitude delineate the locus of all points where the local magnitude at maximum eclipse is equal to a constant value. The maps include curves of constant eclipse magnitude for values of 0.25, 0.5 and 0.75. These curves run exclusively between the curves of maximum eclipse at sunrise and sunset. They are approximately parallel to the northern/southern penumbral limits and the umbral/antumbral paths of central eclipses. The northern and southern limits of the penumbra may be thought of as curves of eclipse magnitude of 0.0. For total eclipses, the northern and southern limits of the umbra are curves of eclipse magnitude of 1.0.

Greatest eclipse is the instant when the axis of the Moon's shadow cone passes closest to Earth's center. Although greatest eclipse differs slightly from the instants of greatest magnitude and greatest duration (for total eclipses), the differences are negligible. The point on Earth's surface intersected by the axis of the Moon's shadow cone at greatest eclipse is marked by an "*". For partial eclipses, the shadow axis misses Earth entirely, so the point of greatest eclipse lies on the day/night terminator and the Sun appears on the horizon.

Data relevant to an eclipse appear in the corners of each map. In the top left corner are the eclipse type (total, hybrid, annular, or partial), the Saros series of the eclipse, and the node of the Moon's orbit where the eclipse occurs. To the top right are the Gregorian calendar date (Julian calendar prior to 1582 Oct 14), the time of greatest eclipse (Terrestrial Dynamical Time), and the value of Delta T (ΔT).

The bottom left corner lists gamma, the minimum distance of the axis of the Moon's shadow cone from Earth's center. The Sun's altitude at the geographic position of greatest eclipse is found to the lower right. The content of the final datum depends on the type of eclipse. If the eclipse is partial then the eclipse magnitude is given. If the eclipse is total, hybrid or annular, then the duration of the total or annular phase is given at the instant of greatest eclipse. A detailed explanation of each of these items appears in the following sections.

4.2 Solar Eclipse Type

There are three types of central solar eclipses:

1. Annular Solar Eclipse – Moon's antumbral shadow traverses Earth; Moon is too far from Earth to completely cover the Sun
2. Total Solar Eclipse – Moon's umbral shadow traverses Earth; Moon is close enough to Earth to completely cover the Sun
3. Hybrid Solar Eclipse – Moon's umbral and antumbral shadows traverse different parts of Earth; eclipse appears either total or annular along different sections of its path

4.3 Saros Series Number

Each eclipse belongs to a Saros series using a numbering system first introduced by van den Bergh (1955). The eclipses with an odd Saros number take place at the ascending node of the Moon's orbit; those with an even Saros number take place at the descending node.

The Saros is a period of 223 synodic months, or approximately 18 years, 11 days and 8 hours. Eclipses separated by this period belong to the same Saros series and share very similar geometry and characteristics.

4.4 Node

A solar eclipse is only possible when New Moon occurs near one of the Moon's two orbital nodes. The ascending node (A. Node) is the point where the Moon travels from south to north through Earth's orbital plane. Similarly, the descending node (D. Node) is the point where the Moon travels from north to south through Earth's orbital plane.

4.5 Calendar Date

The Julian calendar is used prior to 1582 Oct 15. All eclipse dates from 1582 Oct 15 onwards use the modern Gregorian calendar currently found throughout most of the world. Because of the Gregorian Calendar Reform, the day following 1582 Oct 04 (Julian calendar) is 1582 Oct 15 (Gregorian calendar), see Section 6.5.

4.6 Greatest Eclipse

The instant of greatest eclipse occurs when the distance between the axis of the Moon's shadow cone and the center of Earth reaches a minimum. For partial eclipses, the instant of greatest eclipse differs slightly from the instant of greatest magnitude due to Earth's flattening. For total eclipses, the instant of greatest eclipse differs slightly from the instant of greatest duration, although the differences are quite small.

Greatest eclipse is given in Terrestrial Dynamical Time or TD (Section 6.3), which is a time system based on International Atomic Time. As such, TD is the modern equivalent to its predecessor Ephemeris Time and is used in the theories of planetary motion in the Solar System. To determine the geographic visibility of an eclipse, TD is converted to Universal Time (UT1) using the parameter ΔT (Section 6.4).

4.7 ΔT (Delta T)

ΔT (Delta T) is the arithmetic difference, in seconds, between Terrestrial Dynamical Time (TD) and Universal Time (UT1). For more information on ΔT, see Section 6.4.

4.8 Gamma

The quantity gamma is the minimum distance from the axis of the lunar shadow cone to the center of Earth, in units of Earth's equatorial radius. This distance is positive or negative, depending on whether the axis of the shadow cone passes north or south of Earth's center. If gamma is between +0.997 and −0.997, the eclipse is a central one (either total, annular or hybrid). The limiting value 0.997 is due of the flattening of Earth.

4.9 Altitude of Sun

The Sun's altitude at the geographic position intersected by the axis of the lunar shadow cone is given at the instant of greatest eclipse. For partial eclipses, the Sun's altitude is always 0° because the shadow axis misses Earth. In this case, the geographic position corresponds to the point closest to the shadow axis.

4.10 Duration of Central Eclipse

The duration of the total or annular phase (in minutes and seconds) is given at the geographic position intersected by the axis of the lunar shadow cone at the instant of greatest eclipse.

4.11 Eclipse Magnitude

The eclipse magnitude is defined as the fraction of the Sun's diameter occulted by the Moon. For partial eclipses, the eclipse magnitude at the instant of greatest eclipse is given for the geographic position closest to the axis of the Moon's shadow cone. The eclipse magnitude is always less than 1.0 for partial and annular eclipses, but equal to or greater than 1.0 for total and hybrid eclipses.

Section 5: Explanation of Maps in Lower 48 States in Appendix C

5.1 Introduction

The USA experiences 499 central eclipses of the Sun during the 2000-year period from 1001 to 3000. A series of 20 maps (each covering 50 years) centered on the continental USA (i.e., the lower 48 states) illustrates all central paths of those eclipses crossing this region. The maps also show central eclipse paths through southern Canada and northern Mexico.

The paths of total eclipses are shaded gray (light blue in Color Edition) while annular eclipse paths are shaded light gray (pale yellow in Color Edition). Furthermore, total eclipses have a heavy outline while annular eclipses have a light outline. The calendar date of each eclipse appears either above or below the path. The dates of total eclipses are in bold while annular eclipses are in italic. The locations of many major US and Canadian cities are indicated along with state and provenance boundaries.

Theses maps can be used to identify central eclipses visible from a particular city, state or provenance.

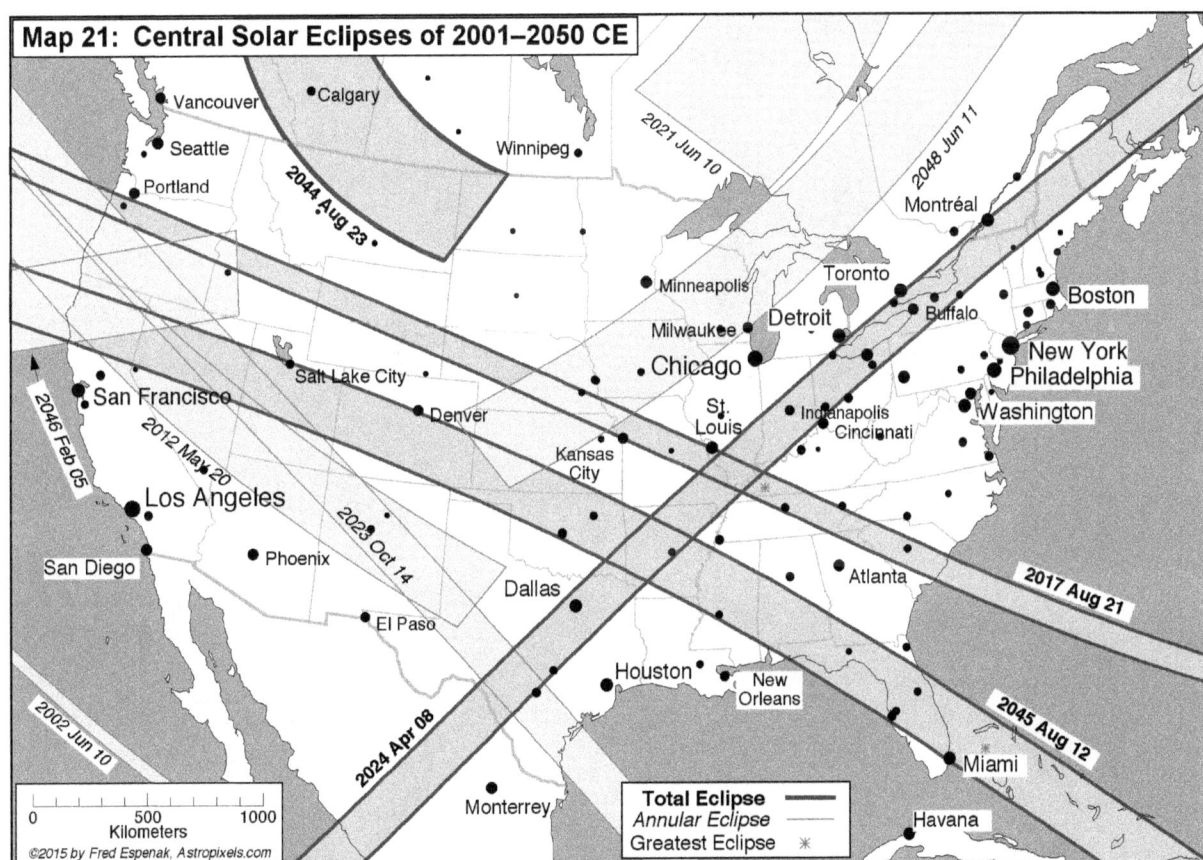

Map 5–1: The map above shows the path of every central solar eclipse in the USA (lower 48 states), southern Canada, and northern Mexico during the 50-year period 2001–2050. The city names shown here are omitted from the maps in Appendix C for readability.

Section 6: Solar Eclipse Predictions

6.1 Mean Lunar Radius

A fundamental parameter used in the prediction of solar eclipses is the Moon's mean radius k, expressed in units of Earth's equatorial radius. This work uses a value of k=0.272281 for all central solar eclipses and represents the mean *minimum* diameter of the Moon. This is smaller than the IAU's adopted a value of k=0.2725076 for the mean *average* lunar radius. The difference is significant because the use of a *minimum* diameter more accurately discriminates between total and annular eclipses. An eclipse is considered total only when the Sun's disk is completely occulted by the Moon. Using a smaller value of k is more representative of the deepest lunar valleys, hence the minimum solid disk radius and ensures that an eclipse is truly total.

6.2 Solar and Lunar Coordinates

The coordinates of the Sun and Moon used in the eclipse predictions presented here have been calculated with the JPL DE406 (Jet Propulsion Laboratory Developmental Ephemeris 406). The DE406 is based upon the International Celestial Reference Frame (ICRF), the adopted reference frame of the International Astronomical Union (IAU). The DE406 is often referred to as the "JPL long ephemeris" because it covers a 6000-year period from -3000 Feb 23 (JED 0625360.50) to +3000 May 06 (JED 02816912.50). Based on the DE405, the accuracy of the interpolating polynomials of the DE406 has been lessened in order to cover the much larger time span. The DE406 does not include nutation or libration.

6.3 Measurement of Time

In 1884, Greenwich Mean Time (GMT) — the mean solar time on the Greenwich Meridian (0° longitude) — was adopted as the standard reference time for clocks around the world. A fundamental basis of GMT is the assumption that Earth's rotation on its axis is constant. By the mid-twentieth century astronomers realized that Earth's rotation is slowing down because of tidal friction with the Moon.

For purposes of orbital calculations, time using Earth's rotation was abandoned for a more uniform time scale based on Earth's orbit about the Sun. In 1952, Ephemeris Time was introduced to address the problem and was used for Solar System ephemeris calculations until 1979.

Terrestrial Dynamical Time (TD) is the modern replacement for Ephemeris Time and is used in the theories of planetary motion in the Solar System. TD is based on International Atomic Time (TAI), which is a high-precision standard using several hundred atomic clocks worldwide. To ensure continuity with Ephemeris Time, TD was defined to match ET for the date 1977 Jan 01.

Civilian time used throughout the world is still based on mean solar time, although indirectly. While Greenwich Mean Time was determined though observations of the Sun, its modern day replacement, Universal Time (actually UT1) is based on Earth's rotation using observations of distant quasars. UT1 is a nonuniform time because Earth is gradually slowing down at an irregular rate.

Coordinated Universal Time (UTC) is derived from International Atomic Time (TAI). UTC was defined to closely parallel UT1. However, the two time systems are intrinsically incompatible since UTC is uniform while UT1 is based on Earth's rotation, which is gradually slowing. In order to keep the two times within 0.9 seconds of each other, a leap second is added to UTC as needed (currently once every few years).

Today, UTC is the time standard used to define time zones around the world. It is the time reference for GPS satellites and aviation, and is used to synchronize the clocks of computers across the Internet.

6.4 ΔT (Delta T)

The orbital positions of the Sun and the Moon, required by eclipse predictions, are calculated using Terrestrial Dynamical Time (TD) because it is a uniform time scale. However, world time zones are based on Universal Time[5] (UT1). In order to convert eclipse predictions from TD to UT1, the difference between these two time scales must be known. The parameter ΔT (delta-T) is the arithmetic difference, in seconds, between the two as:

$$\Delta T = TD - UT1 \qquad\qquad (2\text{–}1)$$

Past values of ΔT can be deduced from historical records. In spite of their relatively low precision, these data represent the only evidence for the value of ΔT prior to 1600. In the centuries following the introduction of the telescope (circa 1609), thousands of high quality observations have been made of lunar occultations of stars, affording valuable data with increased accuracy in the determination of ΔT.

The estimated uncertainty in the value of ΔT is 55 seconds in the year 1000, but it drops to 1 second by 1800. A detailed analysis of historical measurements fitted with cubic splines for ΔT from 1000 to +1950 is presented in Morrison and Stephenson, 2004.

In modern times, the determination of ΔT is made using atomic clocks and radio observations of quasars. From 1955 to 2010, the average 1-year change in ΔT ranges from 0.18 seconds to 1.06 seconds. Future changes in ΔT are unknown since theoretical models of the physical causes are imprecise. Extrapolations from the table weighted by the long period trend from tidal braking of the Moon offer estimates of +71 seconds in 2024, +85 seconds in 2050, and +127 seconds in 2100.

Polynomial expressions for ΔT based on this data can be found at: *eclipsewise.com/help/deltatpoly2014.html*

6.5 Calendar Date

The Gregorian calendar is the civil calendar currently used throughout most of the world. The older Julian calendar was used until 1582 Oct 04. As a consequence of the Gregorian Calendar Reform, the day following 1582 Oct 04 (Julian calendar) is 1582 Oct 15 (Gregorian calendar).

Pope Gregory XIII decreed the use of the Gregorian calendar in 1582 in order to correct a problem in a drift of the seasons. It adopts the convention of a year containing 365 days. Every fourth year is a leap year of 366 days if it is divisible by 4 (e.g., 2004, 2008, etc.). However, whole century years (e.g., 1700, 1800, 1900) are excluded from the leap year rule unless they are also divisible by 400 (e.g., 2000). This dating scheme was designed to keep the vernal equinox on or within a day of March 21. Precession of the equinoxes will eventually produce an error of one day in the Gregorian calendar in about 7700 years.

Prior to the Gregorian Calendar Reform of 1582, the Julian calendar was in wide use. It was less complicated, in that all years divisible by 4 were counted as 366-day leap years, but this simplicity came at a cost. After more than 16 centuries of use, the Julian calendar date of the vernal equinox had drifted 11 days from March 21. It was this failure in the Julian calendar that prompted the Gregorian Calendar Reform.

---

[5] World time zones are actually based on Coordinated Universal Time (UTC). It is an atomic time synchronized and adjusted to stay within a second of astronomically determined Universal Time (UT1) through the addition of an occasional "leap second" to compensate for the gradual slowing of Earth's rotation.

Appendix A

Catalog of Central Solar Eclipses in the USA:
1001 – 3000

Key to Solar Eclipse Catalog

Cat Num — sequential Catalog Number assigned to each eclipse from 1 to 2,389

Canon Plate — plate number assigned to each eclipse map

Calendar Date — Gregorian date of Greatest Eclipse (Julian date prior to 1582 Oct 04)

TD of Greatest Eclipse — Terrestrial Dynamical Time of Greatest Eclipse

ΔT — arithmetic difference between Terrestrial Dynamical Time Universal Time (seconds)

Luna Num — number of synodic months, or lunations, since New Moon on 2000 Jan 06

Saros Num — Saros Series Number of eclipse

Ecl Type — Solar Eclipse Type

      A = Annular Solar Eclipse
      T = Total Solar Eclipse
      H = Hybrid Solar Eclipse

      n = Central eclipse of with no northern limit
      s = Central eclipse of with no southern limit
      + = Non-central eclipse of with no northern limit
      – = Non-central eclipse of with no southern limit
      2 = Hybrid eclipse path begins total and ends annular
      3 = Hybrid eclipse path begins annular and ends total

Gamma — minimum distance from the axis of the lunar shadow to the center of Earth

Ecl Mag — Eclipse Magnitude; fraction of the Sun's diameter obscured by the Moon

Lat & Lng — latitude and longitude where the Sun appears in zenith at greatest eclipse

Sun Alt & Sun Azm — altitude and azimuth of the Sun at greatest eclipse

Path Width — width of the central path (km) at greatest eclipse (total, annular & hybrid eclipses)

Central Line Dur — Central Line Duration (minutes. seconds) at greatest eclipse

USA Geographic Region — Describes locations where central eclipse path passes through
                (L48 = Lower 48 States; AK = Alaska; HI = Hawaii)

Calendar Date	TD of Greatest Eclipse	ΔT s	Luna Num	Saros Num	Ecl Type	Gamma	Ecl Mag	Lat °	Long °	Sun Alt °	Sun Azm °	Path Width km	Central Line Dur	USA Geographic Regions
1001 Mar 27	19:20:51	1551	-11116	113	A	0.9776	0.9637	61.1N	178.0W	11	101	643	02m26s	--- AK
1008 Oct 31	19:08:42	1509	-11022	117	A	0.8677	0.9207	43.0N	90.9W	29	195	602	08m43s	L48 + AK
1011 Aug 31	16:27:45	1494	-10987	109	T	0.1852	1.0612	17.1N	58.0W	79	197	204	05m25s	L48 --
1022 Feb 03	22:00:41	1439	-10858	105	H	0.2142	1.0113	3.3S	145.2W	78	156	40	01m06s	L48 --
1028 Mar 28	23:15:20	1407	-10782	94	A	0.5261	0.9579	35.8N	173.3W	58	159	179	04m24s	L48 --
1035 Nov 02	22:09:27	1369	-10688	98	A	0.6736	0.9352	22.7N	134.1W	48	203	320	07m32s	L48 + AK
1041 Feb 03	20:54:07	1343	-10623	115	T	0.8704	1.0424	40.4N	152.0W	29	149	283	03m13s	--- AK
1042 Jun 20	10:01:08	1336	-10606	92	A	0.9902	0.9959	71.9N	167.9W	7	339	125	00m13s	--- AK
1048 Sep 10	22:09:34	1305	-10529	119	A	0.8531	0.9995	56.8N	119.9W	·31	214	4	00m02s	L48 + AK
1050 Jan 25	20:16:03	1299	-10512	96	T	0.4745	1.0499	10.6N	119.8W	62	170	188	04m41s	L48 --
1051 Jul 10	18:52:18	1292	-10494	111	A	0.3121	0.9957	40.0N	95.1W	72	185	16	00m26s	L48 --
1055 Apr 29	16:40:27	1274	-10447	113	A	0.7928	0.9687	60.8N	105.3W	37	130	183	02m22s	L48 --
1057 Sep 01	20:06:19	1263	-10418	100	A	0.5447	0.9912	34.6N	98.6W	57	215	37	00m46s	L48 + AK
1062 Dec 03	19:13:36	1238	-10353	117	A	0.8342	0.9223	33.8N	105.1W	33	180	534	09m26s	L48 --
1064 Apr 19	13:17:22	1232	-10336	94	A	0.6563	0.9568	52.8N	30.0W	49	157	208	03m58s	L48 --
1076 Mar 07	22:31:47	1178	-10189	105	T	0.1282	1.0290	3.9N	154.1W	83	151	99	02m34s	L48 --
1077 Feb 25	13:47:20	1173	-10177	115	T	0.8215	1.0510	41.0N	49.7W	34	142	290	03m40s	L48 --
1079 Jul 01	13:51:03	1163	-10148	102	T	0.3381	1.0663	42.3N	16.9W	70	195	230	05m12s	L48 --
1082 Apr 30	20:06:11	1151	-10113	94	A	0.7301	0.9556	62.1N	135.6W	43	157	237	03m46s	--- AK
1089 Dec 04	22:34:48	1118	-10019	98	A	0.6962	0.9378	20.5N	147.6W	46	190	320	07m48s	L48 --
1091 May 21	06:14:27	1112	-10001	113	A	0.6408	0.9687	58.7N	69.9E	50	150	146	02m37s	--- AK
1095 Mar 08	22:03:52	1096	-9954	115	T	0.7884	1.0553	41.8N	174.8W	38	140	291	03m54s	--- AK
1097 Jul 11	21:17:14	1086	-9925	102	T	0.4098	1.0667	44.5N	124.6W	66	201	239	05m01s	L48 --
1100 May 11	02:46:31	1074	-9890	94	A	0.8100	0.9537	72.4N	119.9E	36	155	291	03m36s	--- AK
1102 Oct 13	21:59:21	1064	-9860	119	A	0.7649	0.9833	37.8N	133.9W	40	200	91	01m35s	L48 + AK
1104 Feb 27	21:46:53	1059	-9843	96	T	0.5389	1.0568	25.3N	149.3W	57	161	221	04m49s	L48 --
1105 Aug 11	16:12:31	1053	-9825	111	H	0.0982	1.0008	19.0N	56.6W	84	194	3	00m05s	L48 --
1106 Aug 01	04:51:29	1049	-9813	121	T	0.8348	1.0481	70.8N	142.2E	33	214	292	03m00s	--- AK
1111 Oct 04	19:12:27	1028	-9749	100	A	0.6700	0.9712	30.1N	85.5W	48	213	136	02m44s	L48 + AK
1117 Jan 04	19:38:47	1007	-9684	117	A	0.8122	0.9292	31.0N	121.9W	35	164	450	08m19s	L48 --
1125 Dec 26	15:05:28	973	-9573	98	A	0.7026	0.9426	21.8N	40.3W	45	181	298	07m10s	L48 --
1130 Apr 09	22:01:14	957	-9520	105	T	-0.0159	1.0466	9.3N	146.1W	89	331	155	04m04s	L48 --
1137 May 21	16:30:19	930	-9432	104	A	0.2235	0.9504	34.5N	66.0W	77	175	187	05m51s	L48 --
1142 Aug 22	20:14:09	911	-9367	121	T	0.7148	1.0504	52.9N	100.7W	44	207	238	03m36s	L48 + AK
1144 Jan 06	23:19:53	907	-9350	98	A	0.7074	0.9459	23.5N	166.6W	45	176	282	06m36s	L48 --
1149 Apr 09	22:03:59	888	-9285	115	T	0.6480	1.0676	44.9N	171.9W	49	141	287	04m38s	--- AK
1151 Aug 13	20:06:07	880	-9256	102	A	0.6023	1.0635	45.2N	95.7W	53	216	258	04m26s	L48 + AK
1156 Nov 14	22:44:25	862	-9191	119	A	0.7287	0.9687	26.5N	155.2W	43	187	164	03m28s	L48 --
1158 Mar 31	22:07:21	857	-9174	96	T	0.6622	1.0621	46.2N	163.9W	48	155	271	04m33s	--- AK
1165 Nov 05	19:19:13	832	-9080	100	A	0.7378	0.9544	26.7N	92.6W	42	203	242	04m55s	L48 + AK
1171 Feb 06	19:38:26	815	-9015	117	A	0.7619	0.9421	31.6N	129.7W	40	151	321	06m05s	L48 --
1180 Jan 28	15:39:18	786	-8904	98	A	0.7266	0.9542	29.5N	57.2W	43	167	242	05m08s	L48 --
1181 Jul 13	15:11:35	782	-8886	113	A	0.2245	0.9590	33.9N	42.0W	77	186	153	04m42s	L48 --
1191 Jun 23	11:49:21	751	-8763	104	A	0.4832	0.9539	52.0N	13.2E	61	193	193	04m28s	L48 --
1192 Dec 06	15:33:25	747	-8745	119	A	0.7229	0.9614	23.1N	52.9W	44	178	203	04m30s	L48 --
1194 Apr 22	13:44:26	743	-8728	96	T	0.7747	1.0629	62.9N	46.7W	39	151	327	04m03s	L48 --
1196 Sep 23	20:18:43	736	-8698	121	T	0.5822	1.0491	30.9N	112.3W	54	201	199	04m06s	L48 + AK
1198 Feb 07	23:41:15	732	-8681	98	A	0.7437	0.9590	33.9N	178.9E	42	163	221	04m20s	--- + AK
1205 Sep 14	19:58:05	710	-8587	102	T	0.7457	1.0556	41.4N	89.5W	42	220	270	03m55s	L48 --
1209 Jul 03	18:17:39	699	-8540	104	A	0.5691	0.9540	55.8N	77.1W	55	202	204	04m11s	--- + AK
1218 Jul 24	04:55:32	674	-8428	123	A	0.8225	0.9425	72.3N	135.4E	34	209	376	04m34s	--- AK
1219 Dec 08	20:06:04	670	-8411	100	A	0.7660	0.9430	26.3N	112.3W	40	189	327	06m48s	L48 --
1225 Mar 10	18:43:34	656	-8346	117	A	0.6590	0.9596	34.3N	118.9W	49	144	190	03m49s	L48 --
1227 Jul 15	00:51:01	650	-8317	104	A	0.6512	0.9586	58.3N	167.2W	49	212	222	03m59s	--- AK
1230 May 14	04:56:07	643	-8282	96	T	0.9078	1.0597	82.4N	52.5E	24	122	476	03m17s	--- AK
1231 May 03	19:27:01	640	-8270	106	T	0.1762	1.0224	27.8N	112.3W	80	169	78	02m11s	L48 --
1232 Oct 15	13:04:35	636	-8252	121	T	0.5277	1.0469	19.5N	8.9W	58	196	183	04m14s	L48 --
1234 Mar 01	15:26:57	633	-8235	98	A	0.7947	0.9693	45.2N	66.0W	37	155	180	02m49s	L48 --
1243 Mar 22	02:10:22	610	-8123	117	A	0.6105	0.9659	35.5N	129.9E	52	144	152	03m08s	--- AK
1248 May 24	12:24:44	598	-8059	96	T	0.9800	1.0549	78.2N	170.9W	11	13	995	02m42s	--- AK

L48 = Lower 48 States   *AK = Alaska*   *HI = Hawaii*

ATLAS OF CENTRAL SOLAR ECLIPSES IN THE USA

Calendar Date	TD of Greatest Eclipse	ΔT s	Luna Num	Saros Num	Ecl Type	Gamma	Ecl Mag	Lat °	Long °	Sun Alt °	Sun Azm °	Path Width km	Central Line Dur	USA Geographic Regions
1252 Mar 11	23:08:32	588	-8012	98	A	0.8305	0.9745	52.2N	172.8E	34	150	162	02m09s	---- AK
1254 Aug 14	18:19:39	583	-7982	123	A	0.6727	0.9433	52.6N	75.8W	47	204	282	05m23s	---- AK
1257 Jun 13	19:33:18	576	-7947	115	T	0.2409	1.0765	37.6N	112.8W	76	173	255	06m11s	L48 --
1259 Oct 17	21:00:27	571	-7918	102	T	0.8333	1.0464	38.4N	108.8W	33	213	274	03m30s	L48 + AK
1263 Aug 05	14:14:40	562	-7871	104	A	0.8029	0.9516	60.0N	10.2E	36	231	294	03m49s	---- AK
1265 Jan 19	00:57:31	558	-7853	119	A	0.7069	0.9538	23.9N	156.0E	45	159	234	05m08s	---- AK
1274 Jan 09	20:51:20	538	-7742	100	A	0.7885	0.9372	31.3N	132.8W	38	174	380	07m26s	L48 --
1279 Apr 12	16:44:01	527	-7677	117	A	0.4946	0.9788	37.4N	85.3W	60	147	86	01m55s	L48 --
1281 Aug 15	21:07:53	522	-7648	104	A	0.8701	0.9497	59.9N	84.2W	29	239	369	03m50s	L48 --
1285 Jun 04	16:53:55	514	-7601	106	H	0.4023	1.0143	47.2N	70.8W	66	182	54	01m15s	L48 --
1286 Nov 17	15:03:19	511	-7583	121	T	0.4866	1.0441	8.2N	43.9W	61	185	168	04m17s	L48 --
1297 Apr 22	23:51:54	489	-7454	117	A	0.4276	0.9850	37.8N	170.4E	64	150	58	01m22s	L48 --
1299 Aug 27	04:09:34	485	-7425	104	A	0.9310	0.9474	60.0N	179.9E	21	248	525	03m53s	---- AK
1301 Feb 09	17:06:47	482	-7407	119	A	0.6758	0.9533	26.4N	90.3W	47	151	226	04m53s	L48 --
1303 Jun 15	23:53:38	477	-7378	106	H	0.4836	1.0103	52.4N	171.4W	61	189	41	00m52s	---- AK
1308 Sep 15	15:29:59	467	-7313	123	A	0.4965	0.9417	28.2N	42.6W	60	200	247	06m43s	L48 --
1313 Nov 18	23:04:28	457	-7249	102	T	0.8711	1.0395	37.4N	149.3W	29	199	268	03m13s	L48 + AK
1314 Nov 08	14:15:02	455	-7237	112	T	0.2079	1.0244	7.6S	31.3W	78	199	85	02m20s	L48 --
1318 Aug 26	18:27:16	448	-7190	114	H	0.2005	1.0120	17.9N	89.6W	78	209	42	01m06s	L48 --
1319 Feb 21	00:59:41	447	-7184	119	A	0.6517	0.9537	28.0N	150.1E	49	148	218	04m42s	---- AK
1325 Apr 13	17:44:26	436	-7108	108	T	0.2487	1.0551	26.0N	88.8W	75	164	188	04m50s	L48 --
1326 Sep 26	22:53:50	433	-7090	123	A	0.4532	0.9409	21.3N	156.1W	63	199	244	07m07s	---- HI
1330 Jul 16	15:26:55	426	-7043	125	T	0.7308	1.0139	66.5N	35.7W	43	197	70	01m00s	---- AK
1348 Jul 26	22:37:07	395	-6820	125	H	0.6617	1.0098	58.0N	143.3W	48	199	45	00m46s	L48 + AK
1349 Dec 10	16:46:25	393	-6803	102	T	0.8810	1.0371	38.2N	62.7W	28	188	264	03m06s	L48 --
1351 May 25	20:52:01	391	-6785	117	H	0.2016	1.0016	33.7N	135.5W	78	165	6	00m09s	L48 --
1352 May 14	09:04:22	389	-6773	127	A	0.9438	1.0427	73.6N	48.5W	19	81	441	02m18s	---- AK
1355 Mar 14	16:13:53	384	-6738	119	A	0.5792	0.9552	31.2N	79.3W	54	145	196	04m22s	L48 --
1357 Jul 17	20:50:31	381	-6709	106	A	0.7227	0.9942	61.5N	100.5W	43	219	29	00m26s	L48 --
1362 Oct 18	14:09:25	372	-6644	123	A	0.3879	0.9397	9.6N	29.1W	67	194	241	07m48s	L48 --
1364 Mar 04	12:06:37	370	-6627	100	A	0.9094	0.9352	57.9N	28.7W	24	146	579	05m41s	L48 --
1370 May 25	16:28:28	360	-6550	127	T	0.8708	1.0497	76.2N	124.0W	29	117	338	02m51s	---- AK
1372 Sep 27	17:53:13	356	-6521	114	H	0.3304	1.0121	11.9N	80.4W	71	209	44	01m07s	L48 --
1373 Mar 24	23:35:21	356	-6515	119	A	0.5312	0.9561	32.7N	170.9E	58	146	186	04m15s	---- AK
1379 May 16	16:28:57	346	-6439	108	T	0.4396	1.0668	47.0N	70.5W	64	172	243	05m07s	L48 --
1382 Mar 15	19:30:23	342	-6404	100	A	0.9535	0.9344	66.6N	156.5W	17	130	826	05m10s	---- AK
1383 Mar 04	19:25:57	341	-6392	110	A	0.2443	0.9312	10.4N	111.8W	76	162	265	08m56s	L48 --
1384 Aug 17	13:09:04	338	-6374	125	A	0.5355	0.9999	41.7N	4.5W	57	200	1	00m01s	L48 --
1394 Jul 28	14:48:15	324	-6251	116	A	0.1248	0.9501	23.7N	36.5W	83	203	184	05m40s	L48 --
1395 Jan 21	20:05:22	323	-6245	121	T	0.4555	1.0487	7.7N	126.0W	63	160	181	04m21s	L48 --
1397 May 26	23:56:49	320	-6216	108	T	0.5101	1.0692	53.4N	179.9W	59	178	263	05m01s	L48 + AK
1402 Aug 28	20:31:37	312	-6151	125	A	0.4791	0.9943	34.0N	117.4W	61	200	23	00m33s	L48 --
1404 Jan 12	19:20:44	310	-6134	102	T	0.8944	1.0369	43.3N	114.3W	26	171	279	02m58s	L48 --
1424 Jun 26	14:34:24	283	-5881	127	T	0.6425	1.0629	63.1N	36.6W	50	180	270	04m14s	L48 --
1427 Apr 26	20:43:38	279	-5846	119	A	0.3444	0.9583	34.7N	140.2W	70	153	161	04m15s	L48 --
1433 Jun 17	14:48:41	271	-5770	108	T	0.6557	1.0714	64.0N	29.5W	49	196	309	04m38s	L48 --
1437 Apr 05	16:52:04	266	-5723	110	A	0.3973	0.9419	32.1N	79.3W	66	162	233	06m39s	L48 --
1442 Jul 07	21:59:38	260	-5658	127	T	0.5680	1.0654	56.2N	143.3W	55	187	261	04m39s	L48 + AK
1449 Feb 22	21:50:07	252	-5576	121	T	0.4008	1.0561	14.3N	154.6W	66	151	200	04m36s	L48 --
1451 Jun 28	22:15:26	249	-5547	108	T	0.7287	1.0711	67.5N	128.9W	43	210	339	04m23s	L48 + AK
1455 Apr 16	23:43:59	244	-5500	110	A	0.4628	0.9454	40.0N	176.1E	62	162	227	05m53s	L48 --
1458 Feb 13	21:19:37	241	-5465	102	T	0.9373	1.0386	56.3N	162.6W	20	151	375	02m41s	---- AK
1464 May 06	10:46:56	234	-5388	129	A	0.9502	0.9367	71.2N	72.8W	18	83	771	04m17s	---- AK
1466 Sep 09	18:07:52	231	-5359	116	A	0.3965	0.9351	22.2N	80.5W	67	212	260	07m05s	L48 --
1471 Dec 11	22:25:19	225	-5294	133	A	0.9850	0.9871	57.2N	165.0W	9	171	288	01m02s	---- AK
1478 Jul 29	13:01:16	217	-5212	127	T	0.4270	1.0676	41.4N	6.8W	65	194	244	05m18s	L48 --
1480 Dec 01	20:30:36	215	-5183	114	H2	0.4218	1.0155	1.5N	123.9W	65	189	58	01m37s	---- HI
1481 May 28	16:42:57	214	-5177	119	Am	0.1054	0.9577	28.8N	71.9W	84	168	155	04m57s	L48 --
1482 May 17	17:18:59	213	-5165	129	A	0.8682	0.9420	73.4N	137.0W	29	116	434	04m14s	---- AK
1487 Jul 20	13:15:34	208	-5101	108	T	0.8696	1.0673	69.3N	39.6E	29	244	446	03m47s	---- AK

L48 = Lower 48 States   *AK = Alaska*   *HI = Hawaii*

28

Calendar Date	TD of Greatest Eclipse	ΔT s	Luna Num	Saros Num	Ecl Type	Gamma	Ecl Mag	Lat °	Long °	Sun Alt °	Sun Azm °	Path Width km	Central Line Dur	USA Geographic Regions	
1491 May 08	13:11:32	204	-5054	110	A	0.6085	0.9514	56.5N	26.9W	52	166	225	04m30s	L48	--
1496 Aug 08	20:40:13	198	-4989	127	T	0.3626	1.0675	33.9N	122.4W	69	196	236	05m30s	L48	--
1503 Mar 27	22:28:19	191	-4907	121	T	0.2905	1.0640	21.1N	164.1W	73	150	218	05m04s	L48	--
1506 Jul 20	13:46:56	188	-4866	118	T	0.2112	1.0623	30.3N	19.8W	78	202	209	05m08s	L48	--
1509 May 18	19:49:34	185	-4831	110	A	0.6865	0.9539	64.9N	124.3W	46	171	233	03m56s	----	AK
1518 Jun 08	06:15:22	176	-4719	129	A	0.6956	0.9496	67.0N	73.4E	46	162	259	04m13s	----	AK
1520 Oct 11	16:03:18	174	-4690	116	A	0.5277	0.9244	17.8N	49.4W	58	208	329	08m57s	L48	--
1523 Aug 11	04:33:15	172	-4655	108	Tn	0.9968	1.0559	62.7N	136.0W	2	294	-	02m44s	----	AK
1527 May 30	02:22:59	168	-4608	110	A	0.7687	0.9556	73.4N	144.6E	39	180	255	03m28s	L48 + AK	
1531 Mar 18	19:47:20	165	-4561	112	H	0.3817	1.0036	24.3N	122.1W	67	161	13	00m21s	L48	--
1533 Aug 20	05:03:59	163	-4531	137	T	0.9693	1.0479	73.7N	178.3E	13	257	679	02m40s	----	AK
1536 Jun 18	12:43:20	160	-4496	129	A	0.6080	0.9523	61.0N	13.4W	52	174	220	04m17s	L48	--
1543 Feb 03	23:38:50	155	-4414	123	A	0.2736	0.9617	1.0N	177.0W	74	156	143	04m14s	----	HI
1554 Jun 29	19:10:38	146	-4273	129	A	0.5192	0.9546	54.0N	104.8W	58	182	195	04m22s	L48	--
1557 Apr 28	21:59:03	143	-4238	121	A	0.1252	1.0692	24.0N	153.1W	83	157	227	05m42s	L48 + HI	
1558 Apr 18	12:39:26	143	-4226	131	T	0.8930	1.0132	64.1N	67.8W	26	114	100	00m50s	L48	--
1562 Feb 03	17:27:32	140	-4179	133	T	0.9373	1.0091	48.6N	114.5W	20	142	89	00m41s	L48	--
1565 Nov 22	20:49:54	137	-4132	135	A	0.9564	0.9092	51.4N	130.4W	16	184	1221	09m37s	----	AK
1569 Sep 10	20:48:15	135	-4085	137	T	0.8733	1.0428	57.4N	103.4W	29	215	293	02m55s	L48	--
1574 Nov 13	15:12:16	131	-4021	116	A	0.5970	0.9171	14.8N	40.0W	53	197	387	11m03s	L48	--
1576 Apr 28	20:04:43	130	-4003	131	T	0.8329	1.0140	64.8N	168.0W	33	124	86	00m55s	----	AK
1578 Sep 01	20:15:07	129	-3974	118	T	0.4602	1.0408	28.4N	109.6W	62	213	152	03m17s	L48 + AK	
1585 Apr 29	18:28:57	125	-3892	112	H	0.5435	1.0005	46.6N	107.7W	57	162	2	00m03s	L48	--
1597 Mar 17	23:22:38	119	-3745	123	A	0.1879	0.9788	8.4N	173.3W	79	151	77	02m08s	L48	--
1600 Jul 10	12:35:57	117	-3704	120	T	0.2803	1.0238	38.2N	2.7W	74	196	84	02m08s	L48	--
1603 May 11	01:44:58	114	-3669	112	A	0.6107	0.9987	54.7N	142.6E	52	163	6	00m07s	L48 + AK	
1608 Aug 10	15:00:05	108	-3604	129	A	0.2722	0.9581	31.0N	39.6W	74	194	158	04m46s	L48	--
1618 Jul 21	19:44:29	95	-3481	120	T	0.3558	1.0260	40.4N	106.3W	69	201	94	02m13s	L48	--
1619 Jan 15	20:38:06	95	-3475	125	A	0.2349	0.9422	8.1S	130.4W	76	165	220	07m16s	L48	--
1620 Jan 04	20:51:03	93	-3463	135	A	0.9322	0.9081	45.0N	146.5W	21	165	976	10m13s	----	AK
1623 Oct 23	21:17:09	88	-3416	137	T	0.7770	1.0298	37.8N	128.0W	39	199	159	02m31s	L48 + AK	
1625 Mar 08	17:32:38	86	-3399	114	T	0.4965	1.0434	23.9N	89.4W	60	161	166	03m50s	L48	--
1627 Aug 11	04:17:13	82	-3369	139	H	0.9401	1.0001	77.7N	173.2W	19	253	1	00m00s	----	AK
1628 Dec 25	15:08:46	80	-3352	116	A	0.6264	0.9153	15.4N	44.0W	51	184	413	12m02s	L48	--
1632 Oct 13	20:09:38	74	-3305	118	A	0.5872	1.0220	23.7N	108.2W	54	210	91	01m55s	L48 + AK	
1647 Jan 05	23:10:59	53	-3129	116	A	0.6336	0.9161	16.9N	166.5W	51	179	413	11m50s	L48 + HI	
1648 Jun 21	00:43:22	51	-3111	131	H	0.5483	1.0102	56.7N	164.0E	56	171	42	00m49s	L48 + AK	
1651 Apr 19	22:04:36	47	-3076	123	A	0.0433	0.9976	13.7N	152.4W	87	154	8	00m14s	L48	--
1656 Jan 26	12:48:10	40	-3017	135	A	0.9122	0.9106	43.2N	34.1W	24	154	820	09m38s	L48	--
1659 Nov 14	14:10:07	35	-2970	137	T	0.7432	1.0208	29.2N	28.2W	42	190	106	01m56s	L48	--
1670 Apr 19	18:12:20	23	-2841	133	T	0.7191	1.0476	50.6N	123.3W	44	137	225	03m15s	L48	--
1672 Aug 22	17:44:06	20	-2812	120	T	0.5593	1.0288	41.2N	66.2W	56	215	117	02m15s	L48 + AK	
1673 Feb 16	20:49:18	20	-2806	125	A	0.1951	0.9409	1.8S	133.5W	79	154	223	06m52s	L48	--
1674 Feb 05	20:41:35	19	-2794	135	A	0.8979	0.9129	42.8N	155.7W	26	149	736	09m09s	----	AK
1675 Jun 23	05:44:38	18	-2777	112	A	0.9218	0.9835	84.1N	166.1W	22	282	154	01m01s	----	AK
1677 Nov 24	22:44:03	16	-2747	137	T	0.7332	1.0166	26.3N	159.6W	43	186	84	01m36s	L48	--
1679 Apr 10	17:55:13	14	-2730	114	T	0.6070	1.0565	43.8N	102.2W	52	157	233	04m17s	L48 + HI	
1680 Sep 22	19:06:23	13	-2712	129	A	0.0161	0.9578	0.7N	108.2W	89	198	153	05m08s	----	HI
1683 Jan 27	15:10:08	12	-2683	116	A	0.6525	0.9195	22.1N	50.6W	49	171	401	10m44s	L48	--
1684 Jul 12	14:40:34	11	-2665	131	H	0.3927	1.0041	45.2N	37.1W	67	184	16	00m23s	L48	--
1686 Nov 15	21:04:59	10	-2636	118	H	0.6578	1.0048	20.2N	126.0W	49	200	22	00m28s	----	AK
1688 Apr 30	01:57:33	9	-2618	133	T	0.6621	1.0535	51.4N	124.4E	48	141	234	03m40s	----	AK
1690 Sep 03	01:17:46	8	-2589	120	T	0.6173	1.0287	40.3N	177.4W	52	217	122	02m13s	----	HI
1694 Jun 22	16:08:45	8	-2542	122	A	0.2556	0.9517	38.4N	59.7W	75	187	183	05m27s	L48	--
1697 Apr 21	01:49:21	8	-2507	114	T	0.6559	1.0602	51.4N	136.9E	49	157	262	04m18s	----	AK
1701 Feb 07	23:04:53	8	-2460	116	A	0.6662	0.9219	25.9N	171.7W	48	167	393	09m55s	L48	--
1713 Dec 17	16:04:20	9	-2301	137	H	0.7249	1.0094	23.1N	64.8W	43	176	47	00m56s	L48	--
1717 Oct 04	18:08:27	9	-2254	139	H	0.6563	1.0104	34.6N	81.1W	49	201	47	00m56s	L48 + AK	
1722 Dec 08	14:07:34	10	-2190	118	A	0.6808	0.9955	19.5N	25.4W	47	191	21	00m28s	L48	--
1724 May 22	17:10:08	10	-2172	133	T	0.5319	1.0640	50.8N	92.9W	58	154	247	04m33s	L48	--

L48 = Lower 48 States   AK = Alaska   HI = Hawaii

Calendar Date	TD of Greatest Eclipse	ΔT s	Luna Num	Saros Num	Ecl Type	Gamma	Ecl Mag	Lat °	Long °	Sun Alt °	Sun Azm °	Path Width km	Central Line Dur	USA Geographic Regions	
1727 Mar 22	19:47:55	10	-2137	125	A	0.0996	0.9432	5.7N	118.0W	84	151	211	06m20s	L48	--
1728 Mar 10	19:38:55	10	-2125	135	A	0.8172	0.9233	42.8N	144.6W	35	139	485	07m25s	----	AK
1733 May 13	17:18:28	11	-2061	114	T	0.7712	1.0656	67.9N	99.5W	39	157	339	04m06s	----	AK
1735 Oct 16	02:10:33	11	-2031	139	H	0.6202	1.0110	28.3N	155.2E	51	198	48	01m02s	----	HI
1737 Mar 01	14:35:17	11	-2014	116	A	0.7098	0.9283	36.0N	50.1W	45	160	378	08m04s	L48	--
1740 Dec 18	22:43:17	11	-1967	118	A	0.6876	0.9917	19.9N	156.4W	46	187	40	00m53s	L48 + HI	
1742 Jun 03	00:39:56	12	-1949	133	T	0.4607	1.0683	49.0N	160.2E	62	161	251	05m00s	L48 + AK	
1744 Oct 06	00:51:23	12	-1920	120	T	0.7521	1.0263	37.0N	169.1W	41	216	132	02m04s	----	AK
1752 May 13	17:56:28	13	-1826	124	T	0.1090	1.0637	24.9N	91.1W	84	171	210	05m42s	L48	--
1755 Mar 12	22:09:32	13	-1791	116	A	0.7413	0.9319	42.2N	167.4W	42	156	375	07m07s	----	AK
1756 Aug 25	18:46:17	14	-1773	131	Am	0.1009	0.9853	16.1N	99.5W	84	196	52	01m38s	----	HI
1757 Aug 14	22:16:45	14	-1761	141	A	0.8808	0.9407	71.6N	113.4W	28	224	467	04m36s	L48	--
1766 Aug 05	17:56:57	15	-1650	122	A	0.6023	0.9433	50.2N	67.0W	53	214	260	05m15s	----	AK
1768 Jan 19	18:09:29	15	-1632	137	H	0.7195	1.0022	23.9N	103.2W	44	162	11	00m13s	L48	--
1777 Jan 09	15:55:35	16	-1521	118	A	0.6987	0.9859	22.4N	58.9W	46	177	70	01m32s	L48	--
1778 Jun 24	15:34:55	16	-1503	133	T	0.3127	1.0746	41.8N	55.0W	72	175	255	05m52s	L48	--
1780 Oct 27	17:18:27	17	-1474	120	T	0.8083	1.0244	35.6N	58.6W	36	210	138	02m00s	L48	--
1782 Apr 12	17:24:47	17	-1456	135	A	0.6745	0.9370	45.1N	107.1W	47	140	311	05m51s	L48	--
1784 Aug 16	00:31:53	17	-1427	122	A	0.6819	0.9402	50.9N	159.8W	47	220	299	05m23s	----	AK
1791 Apr 03	12:55:13	16	-1345	116	A	0.8236	0.9394	57.1N	39.5W	34	147	394	05m21s	L48	--
1799 May 05	00:13:07	14	-1245	125	A	-0.1310	0.9476	9.3N	178.9E	83	338	194	06m20s	----	HI
1800 Apr 24	00:24:00	13	-1233	135	A	0.6125	0.9417	45.7N	151.3E	52	143	269	05m27s	----	AK
1803 Feb 21	21:18:46	12	-1198	127	T	-0.0075	1.0492	11.1S	135.9W	90	337	163	04m09s	L48	--
1806 Jun 16	16:24:27	12	-1157	124	T	0.3203	1.0604	42.2N	64.6W	71	184	210	04m55s	L48	--
1809 Apr 14	20:07:11	12	-1122	116	A	0.8742	0.9429	65.8N	157.3W	29	139	435	04m35s	----	AK
1811 Sep 17	18:43:45	12	-1092	141	A	0.6798	0.9345	43.0N	85.9W	47	204	330	06m51s	L48 + AK	
1821 Aug 27	15:19:42	11	-969	132	A	0.0671	0.9661	13.6N	47.8W	86	207	123	03m38s	L48	--
1822 Feb 21	19:40:40	11	-963	137	A	0.6914	0.9996	28.6N	132.3W	46	150	2	00m02s	L48	--
1825 Dec 09	20:21:45	9	-916	139	H2	0.5296	1.0148	9.2N	127.4W	58	180	60	01m34s	L48	--
1829 Sep 28	01:46:53	8	-869	141	A	0.6244	0.9318	34.9N	164.3E	51	202	323	07m43s	----	HI
1831 Feb 12	17:21:44	7	-852	118	A	0.7288	0.9807	31.9N	88.3W	43	165	100	01m57s	L48	--
1834 Nov 30	18:56:35	6	-805	120	T	0.8497	1.0233	34.9N	91.6W	32	197	150	02m02s	L48	--
1838 Sep 18	20:55:56	5	-758	122	A	0.8868	0.9289	52.4N	90.6W	27	232	562	06m06s	L48	--
1839 Sep 07	22:23:26	5	-746	132	Am	0.1324	0.9661	12.8N	152.7W	82	209	123	03m34s	----	HI
1849 Feb 23	01:38:09	7	-629	118	A	0.7474	0.9796	36.7N	144.3E	41	161	108	01m58s	----	AK
1850 Aug 07	21:33:54	7	-611	133	T	0.0215	1.0769	17.7N	141.8W	89	191	249	06m50s	----	HI
1851 Jul 28	14:33:42	7	-599	143	T	0.7644	1.0577	68.0N	19.6W	40	201	296	03m41s	----	AK
1854 May 26	20:42:53	7	-564	135	A	0.3918	0.9551	43.3N	140.1W	67	159	178	04m32s	L48	--
1856 Sep 29	03:59:44	7	-535	122	A	0.9420	0.9246	54.3N	169.1E	19	236	831	06m21s	----	AK
1860 Jul 18	14:26:24	8	-488	124	T	0.5487	1.0500	52.5N	20.3W	56	205	198	03m39s	L48	--
1865 Oct 19	16:21:13	5	-423	141	A	0.5366	0.9263	21.3N	60.2W	57	196	326	09m27s	L48	--
1869 Aug 07	22:01:05	1	-376	143	T	0.6960	1.0551	59.1N	133.2W	46	202	254	03m48s	L48 + AK	
1875 Sep 29	12:58:09	-4	-300	132	A	0.2427	0.9656	10.0N	10.1W	76	209	127	03m36s	L48	--
1876 Mar 25	20:05:06	-4	-294	137	A	0.6143	0.9999	34.8N	141.1W	52	144	1	00m01s	----	HI
1878 Jul 29	21:47:18	-5	-265	124	T	0.6232	1.0450	53.8N	124.0W	51	213	191	03m11s	L48 + AK	
1880 Jan 11	22:34:25	-5	-247	139	T	0.5136	1.0212	8.3N	164.1W	59	166	84	02m07s	L48	--
1885 Mar 16	17:45:42	-6	-183	118	A	0.8030	0.9778	48.9N	106.1W	36	153	132	01m55s	L48	--
1889 Jan 01	21:16:50	-6	-136	120	T	0.8603	1.0262	36.7N	137.6W	30	181	175	02m17s	L48 + AK	
1894 Apr 06	03:53:41	-6	-71	137	H	0.5740	1.0001	36.7N	102.4E	55	144	1	00m01s	----	AK
1900 May 28	14:53:56	-2	5	126	T	0.3943	1.0249	44.8N	46.5W	67	175	92	02m10s	L48	--
1908 Jun 28	16:29:51	8	105	135	A	0.1390	0.9655	31.4N	67.2W	82	177	126	04m00s	L48	--
1918 Jun 08	22:07:43	20	228	126	T	0.4658	1.0292	50.9N	152.0W	62	180	112	02m23s	L48	--
1919 Nov 22	15:14:12	21	246	141	A	0.4549	0.9198	6.9N	48.9W	63	186	341	11m37s	L48	--
1923 Sep 10	20:47:29	23	293	143	T	0.5149	1.0430	34.7N	121.8W	59	201	167	03m37s	L48	--
1925 Jan 24	14:54:03	24	310	120	T	0.8661	1.0304	40.5N	49.6W	30	170	206	02m32s	L48	--
1927 Jun 29	06:23:27	24	340	145	T	0.8163	1.0128	78.1N	73.8E	35	167	77	00m50s	----	AK
1930 Apr 28	19:03:34	24	375	137	H	0.4731	1.0003	39.4N	121.2W	62	149	1	00m01s	L48	--
1932 Aug 31	20:03:41	24	404	124	T	0.8307	1.0257	54.5N	79.5W	34	232	155	01m45s	L48	--
1939 Apr 19	16:45:53	24	486	118	A	0.9388	0.9731	73.1N	129.1W	20	118	285	01m49s	----	AK
1940 Apr 07	20:21:21	24	498	128	A	0.2190	0.9394	19.2N	128.5W	77	163	230	07m30s	L48	--

L48 = Lower 48 States   AK = Alaska   HI = Hawaii

Calendar Date	TD of Greatest Eclipse	ΔT s	Luna Num	Saros Num	Ecl Type	Gamma	Ecl Mag	Lat °	Long °	Sun Alt °	Sun Azm °	Path Width km	Central Line Dur	USA Geographic Regions
1943 Feb 04	23:38:10	26	533	120	T	0.8734	1.0331	43.6N	175.1E	29	165	229	02m39s	---- AK
1945 Jul 09	13:27:45	27	563	145	T	0.7356	1.0180	70.0N	17.2W	42	184	92	01m15s	L48 --
1948 May 09	02:26:04	28	598	137	A	0.4133	0.9999	39.8N	131.2E	65	153	0	00m00s	---- AK
1950 Sep 12	03:38:47	29	627	124	T	0.8903	1.0182	54.8N	172.3E	27	236	134	01m14s	---- AK
1951 Sep 01	12:51:51	30	639	134	A	0.1557	0.9747	16.5N	8.5W	81	208	91	02m36s	L48 --
1954 Jun 30	12:32:38	31	674	126	T	0.6134	1.0357	60.5N	4.2E	52	197	153	02m35s	L48 --
1959 Oct 02	12:27:00	33	739	143	T	0.4207	1.0325	20.4N	1.4W	65	199	120	03m02s	L48 --
1963 Jul 20	20:36:13	35	786	145	T	0.6571	1.0224	61.7N	119.6W	49	191	101	01m40s	L48 + AK
1970 Mar 07	17:38:30	40	868	139	T	0.4473	1.0414	18.2N	94.7W	63	150	153	03m28s	L48 --
1972 Jul 10	19:46:38	43	897	126	T	0.6872	1.0379	63.5N	94.2W	46	209	175	02m36s	---- AK
1979 Feb 26	16:55:06	50	979	120	T	0.8981	1.0391	52.1N	94.5W	26	153	298	02m49s	L48 --
1984 May 30	16:45:42	54	1044	137	A	0.2755	0.9980	37.5N	76.7W	74	163	7	00m11s	L48 --
1990 Jul 22	03:03:07	57	1120	126	T	0.7597	1.0391	65.2N	168.9E	40	222	201	02m33s	---- AK
1991 Jul 11	19:07:01	58	1132	136	Tm	-0.0041	1.0800	22.0N	105.2W	90	30	258	06m53s	---- HI
1992 Jan 04	23:05:37	58	1138	141	A	0.4091	0.9179	1.0N	169.7W	66	169	340	11m41s	L48 --
1994 May 10	17:12:27	60	1167	128	A	0.4077	0.9431	41.5N	84.1W	66	168	230	06m13s	L48 --
2012 May 20	23:53:54	66	1390	128	A	0.4828	0.9439	49.1N	176.3E	61	171	237	05m46s	L48 + AK
2017 Aug 21	18:26:40	69	1455	145	T	0.4367	1.0306	37.0N	87.7W	64	198	115	02m40s	L48 --
2023 Oct 14	18:00:41	71	1531	134	A	0.3753	0.9520	11.4N	83.1W	68	208	187	05m17s	L48 --
2024 Apr 08	18:18:29	71	1537	139	T	0.3431	1.0566	25.3N	104.1W	70	149	198	04m28s	L48 --
2033 Mar 30	18:02:36	75	1648	120	T	0.9778	1.0462	71.3N	155.8W	11	111	781	02m37s	---- AK
2039 Jun 21	17:12:54	79	1725	147	A	0.8312	0.9454	78.9N	102.1W	33	153	365	04m05s	---- AK
2044 Aug 23	01:17:02	82	1789	126	T	0.9613	1.0364	64.3N	120.5W	15	264	453	02m04s	L48 --
2045 Aug 12	17:42:39	82	1801	136	T	0.2116	1.0774	25.9N	78.6W	78	206	256	06m06s	L48 --
2046 Feb 05	23:06:26	83	1807	141	A	0.3765	0.9232	4.8N	171.4W	68	157	310	09m42s	L48 --
2048 Jun 11	12:58:53	84	1836	128	A	0.6468	0.9441	63.7N	11.5W	49	184	271	04m58s	L48 --
2052 Mar 30	18:31:53	87	1883	130	T	0.3239	1.0466	22.4N	102.6W	71	161	164	04m08s	L48 --
2056 Jan 16	22:16:45	89	1930	132	A	0.4199	0.9760	3.9N	153.6W	65	175	95	02m52s	L48 --
2057 Jul 01	23:40:15	90	1948	147	A	0.7455	0.9464	71.5N	176.3W	41	177	298	04m22s	---- AK
2066 Jun 22	19:25:48	97	2059	128	A	0.7330	0.9435	70.1N	96.5W	43	198	309	04m40s	---- AK
2077 Nov 15	17:07:56	106	2200	134	A	0.4705	0.9371	7.8N	71.0W	62	199	262	07m54s	L48 --
2078 May 11	17:56:55	106	2206	139	T	0.1838	1.0701	28.1N	93.9W	79	158	232	05m40s	L48 --
2079 May 01	10:50:13	107	2218	149	T	0.9081	1.0512	66.2N	46.5W	24	108	406	02m55s	L48 --
2084 Jul 03	01:50:26	112	2282	128	A	0.8208	0.9421	75.0N	169.3W	35	222	377	04m25s	L48 + AK
2093 Jul 23	12:32:04	120	2394	147	A	0.5717	0.9463	54.6N	1.1E	55	191	241	05m11s	L48 --
2097 May 11	18:34:31	124	2441	149	T	0.8516	1.0538	67.4N	149.8W	31	121	339	03m10s	---- AK
2099 Sep 14	16:57:53	127	2470	136	T	0.3942	1.0684	23.4N	63.1W	67	211	241	05m18s	L48 --
2100 Mar 10	22:28:11	127	2476	141	A	0.3077	0.9338	12.0N	162.7W	72	151	257	07m29s	L48 + HI
2106 May 03	18:19:19	134	2552	130	T	0.4681	1.0472	43.1N	102.6W	62	164	178	03m47s	L48 --
2108 Oct 05	01:01:20	136	2582	155	T	0.8722	1.0551	52.5N	172.3W	29	209	371	03m50s	---- + AK
2110 Feb 18	23:31:35	138	2599	132	A	0.4438	0.9888	14.1N	175.6W	64	165	44	01m12s	L48 --
2111 Aug 04	19:00:22	139	2617	147	A	0.4867	0.9455	46.0N	95.7W	61	194	230	05m42s	L48 --
2121 Jul 14	16:42:38	151	2740	138	A	0.2125	0.9758	33.6N	64.7W	78	197	88	02m32s	L48 --
2122 Dec 28	22:00:56	153	2758	153	A+	1.0072	0.9451	65.3N	170.2W	0	161	-	------	---- AK
2124 May 14	01:59:10	155	2775	130	T	0.5286	1.0464	50.3N	142.7E	58	167	182	03m34s	---- AK
2131 Dec 19	17:06:50	164	2869	134	A	0.5165	0.9267	7.6N	73.2W	59	186	321	10m14s	L48 --
2137 Mar 21	18:16:38	171	2934	151	A	0.9369	0.9769	55.6N	145.3W	20	121	233	01m40s	---- AK
2144 Oct 26	17:32:40	181	3028	155	T	0.8037	1.0512	39.2N	71.7W	36	198	284	04m05s	L48 --
2149 Dec 30	01:13:04	189	3092	134	A	0.5253	0.9245	8.6N	164.2E	58	182	334	10m42s	---- HI
2153 Oct 17	17:12:17	194	3139	136	T	0.5259	1.0560	18.8N	66.3W	58	208	214	04m36s	L48 + AK
2154 Apr 12	20:43:01	195	3145	141	A	0.1794	0.9478	18.2N	134.8W	80	152	195	05m42s	L48 --
2164 Mar 23	00:02:47	210	3268	132	H	0.5096	1.0051	30.4N	171.4E	59	159	20	00m29s	L48 --
2165 Sep 05	14:52:45	212	3286	147	A	0.2549	0.9406	20.7N	38.1W	75	198	227	07m22s	L48 --
2169 Jun 25	00:37:09	218	3333	149	T	0.5841	1.0562	59.2N	167.9E	54	173	229	03m58s	L48 + AK
2178 Jun 16	00:20:42	233	3444	130	T	0.7379	1.0396	71.0N	175.9W	42	190	198	02m36s	L48 + AK
2183 Sep 16	21:42:37	241	3509	147	A	0.1877	0.9384	12.8N	142.6W	79	198	233	07m53s	---- HI
2193 Aug 26	20:09:19	259	3632	138	Am	0.5200	0.9806	37.4N	103.6W	58	214	80	01m45s	L48 + AK
2197 Jun 15	17:59:33	265	3679	140	A	0.0574	0.9864	26.8N	88.3W	87	184	48	01m32s	L48 --
2198 Nov 28	19:12:46	268	3697	155	T	0.7459	1.0442	26.9N	106.7W	42	184	221	03m58s	L48 --
2200 Apr 14	15:49:57	271	3714	132	T	0.5847	1.0165	43.8N	69.1W	54	158	69	01m23s	L48 --

L48 = Lower 48 States   AK = Alaska   HI = Hawaii

ATLAS OF CENTRAL SOLAR ECLIPSES IN THE USA

Calendar Date	TD of Greatest Eclipse	ΔT s	Luna Num	Saros Num	Ecl Type	Gamma	Ecl Mag	Lat °	Long °	Sun Alt °	Sun Azm °	Path Width km	Central Line Dur	USA Geographic Regions	
2205 Jul 17	15:17:59	280	3779	149	T	0.4366	1.0525	47.2N	43.7W	64	186	193	04m10s	L48	--
2207 Nov 20	18:30:26	285	3808	136	T	0.6028	1.0434	15.8N	88.5W	53	198	180	03m56s	L48	--
2213 Feb 21	14:30:14	295	3873	153	A	0.9635	0.9230	53.4N	79.4W	15	133	1080	06m44s	L48	--
2214 Jul 08	14:52:45	297	3890	130	T	0.8926	1.0303	78.1N	27.5E	26	253	230	01m46s	----	AK
2218 Apr 25	23:33:14	305	3937	132	T	0.6321	1.0219	51.1N	173.6E	51	158	96	01m43s	----	AK
2226 May 27	00:45:11	321	4037	141	A	-0.0810	0.9670	16.8N	170.7E	85	344	119	03m55s	----	HI
2238 Oct 08	21:01:18	347	4190	157	A	0.7459	0.9618	40.1N	120.5W	41	202	206	03m47s	L48 + AK
2240 Feb 23	17:14:11	350	4207	134	T	0.5859	1.0228	24.7N	83.8W	54	163	356	09m41s	L48	--
2245 May 26	15:42:04	361	4272	151	T	0.6089	1.0201	56.7N	72.2W	52	153	86	01m30s	L48	--
2247 Sep 29	18:01:05	366	4301	138	A	0.6961	0.9801	35.6N	66.7W	46	216	96	01m47s	L48	--
2251 Jul 19	14:18:46	375	4348	140	A	0.3062	0.9773	38.0N	25.0W	72	200	85	02m16s	L48	--
2252 Dec 31	21:37:06	378	4366	155	T	0.7257	1.0389	23.1N	149.9W	43	170	189	03m33s	L48	HI
2254 May 17	14:43:39	381	4383	132	T	0.7426	1.0315	66.7N	54.9W	42	161	160	02m09s	L48	--
2258 Mar 06	00:58:23	390	4430	134	A	0.6101	0.9239	30.2N	157.9E	52	160	358	09m04s	----	AK
2259 Aug 19	13:22:17	393	4448	149	T	0.2226	1.0412	25.3N	14.5W	77	195	141	03m49s	L48	--
2261 Dec 22	20:38:49	398	4477	136	T	0.6360	1.0337	16.1N	125.0W	50	185	147	03m17s	L48	--
2263 Jun 06	22:58:57	402	4495	151	T	0.5365	1.0261	54.4N	174.0W	57	162	105	02m01s	L48 + AK
2265 Oct 10	01:37:34	407	4524	138	A	0.7405	0.9796	35.1N	179.3E	42	215	105	01m51s	----	HI
2267 Mar 26	13:33:45	411	4542	153	A	0.8810	0.9290	52.3N	64.5W	28	128	549	06m03s	L48	--
2269 Jul 29	21:03:04	416	4571	140	A	0.3893	0.9732	39.9N	122.1W	67	205	104	02m35s	L48	--
2285 Apr 05	20:55:22	455	4765	153	A	0.8378	0.9315	52.9N	172.3W	33	129	459	05m50s	----	AK
2290 Jun 08	05:35:48	468	4829	132	T	0.8713	1.0382	83.8N	100.0E	29	182	265	02m14s	----	AK
2292 Nov 09	20:20:06	474	4859	157	A	0.6375	0.9635	22.0N	120.0W	50	191	171	04m14s	L48	--
2294 Mar 27	16:02:22	477	4876	134	A	0.6777	0.9269	43.2N	73.5W	47	156	370	07m42s	L48	--
2298 Jan 13	14:16:26	487	4923	136	T	0.6474	1.0296	19.0N	32.9W	50	176	131	02m52s	L48	--
2301 Nov 01	17:19:32	497	4970	138	A	0.8080	0.9786	34.8N	58.1W	36	209	126	02m01s	L48	--
2307 Feb 04	00:08:00	511	5035	155	T	0.7125	1.0373	25.7N	165.9E	44	156	176	03m12s	----	AK
2308 Jun 19	12:57:52	515	5052	132	T	0.9402	1.0396	84.1N	119.7E	19	313	401	02m08s	----	AK
2312 Apr 07	23:19:31	525	5099	134	A	0.7231	0.9286	50.8N	173.7E	43	153	385	07m00s	----	AK
2314 Sep 10	20:49:10	532	5129	159	A	0.8247	0.9654	56.8N	104.3W	34	212	220	02m54s	L48 + AK
2316 Jan 25	23:05:16	535	5146	136	T	0.6527	1.0282	21.4N	166.9W	49	172	126	02m42s	L48	--
2317 Jul 09	20:42:40	539	5164	151	T	0.3078	1.0406	40.4N	126.2W	72	182	143	03m32s	L48	--
2323 Sep 01	17:26:08	556	5240	140	A	0.6253	0.9584	41.7N	56.2W	51	218	191	03m48s	----	AK
2324 Aug 20	18:28:21	559	5252	150	A	-0.1260	0.9449	5.7N	97.0W	83	25	205	06m33s	----	HI
2339 May 09	18:08:03	601	5434	153	A	0.6672	0.9392	54.7N	115.5W	48	143	300	05m24s	L48 + HI
2341 Sep 12	00:22:46	608	5463	140	A	0.6950	0.9529	41.7N	157.4W	46	220	234	04m19s	----	AK
2343 Feb 25	17:32:17	613	5481	155	T	0.6913	1.0385	29.6N	98.7W	46	149	175	03m06s	L48	--
2345 Jun 30	20:26:17	620	5510	142	T	0.3267	1.0797	42.1N	118.7W	71	192	272	06m07s	L48	--
2346 Dec 13	20:55:35	624	5528	157	A	0.5848	0.9665	12.8N	134.1W	54	178	149	04m04s	L48	--
2348 Apr 29	13:29:00	628	5545	134	A	0.8338	0.9315	68.1N	49.9W	33	145	466	05m40s	L48	--
2352 Feb 16	16:32:05	639	5592	136	T	0.6709	1.0266	28.5N	72.8W	48	164	121	02m25s	L48	--
2354 Jul 21	03:28:21	647	5622	161	T	0.8870	1.0697	81.4N	170.6E	27	221	499	03m51s	----	AK
2355 Dec 04	17:58:36	651	5639	138	A	0.8609	0.9792	36.0N	75.4W	30	195	145	02m02s	L48	--
2357 May 20	00:54:22	656	5657	153	A	0.5960	0.9415	53.9N	149.9E	53	150	268	05m24s	----	AK
2361 Mar 08	02:05:55	667	5704	155	T	0.6742	1.0396	31.9N	131.7E	47	146	176	03m06s	----	AK
2367 Apr 29	21:30:02	686	5780	144	Am	0.1452	0.9607	22.8N	142.2W	82	167	144	04m38s	L48	--
2368 Oct 12	18:37:19	691	5798	159	A	0.6672	0.9522	32.5N	86.9W	48	199	233	05m13s	L48 + AK
2370 Feb 27	01:07:02	695	5815	136	T	0.6865	1.0262	33.2N	156.0E	46	161	122	02m17s	----	AK
2378 Sep 22	14:45:47	723	5921	150	A	0.0905	0.9396	4.8N	37.6W	85	209	225	06m54s	L48	--
2381 Jul 22	11:25:01	732	5956	142	T	0.4749	1.0777	46.9N	25.9E	61	205	285	05m32s	L48	--
2386 Oct 24	02:06:42	749	6021	159	A	0.6267	0.9475	26.1N	157.7E	51	196	246	06m09s	----	HI
2390 Aug 11	18:31:26	762	6068	161	T	0.7440	1.0724	61.3N	73.6W	42	206	353	04m41s	----	AK
2393 Jun 10	14:08:40	771	6103	153	A	0.4389	0.9453	48.7N	35.5W	64	166	224	05m34s	L48	--
2395 Oct 14	21:49:15	779	6132	140	A	0.8692	0.9354	44.0N	113.5W	29	219	471	06m07s	L48	--
2396 Oct 02	21:48:06	783	6144	150	A	0.1494	0.9375	3.5N	142.3W	81	209	234	07m12s	----	HI
2397 Mar 29	18:49:51	784	6150	155	T	0.6221	1.0423	36.7N	119.9W	51	144	178	03m11s	L48	--
2399 Aug 02	18:55:13	792	6179	142	T	0.5482	1.0754	48.0N	81.5W	57	211	291	05m14s	L48 + AK
2401 Jan 14	22:15:19	797	6197	157	A	0.5617	0.9735	11.9N	157.5W	56	165	114	03m00s	L48	--
2406 Mar 20	17:57:22	815	6261	136	T	0.7327	1.0258	44.6N	102.4W	43	155	128	02m03s	L48	--
2410 Jan 05	19:31:38	828	6308	138	A	0.8749	0.9842	38.8N	109.3W	29	179	116	01m31s	L48	--

L48 = Lower 48 States   AK = Alaska   HI = Hawaii

32

Calendar Date	TD of Greatest Eclipse	ΔT s	Luna Num	Saros Num	Ecl Type	Gamma	Ecl Mag	Lat °	Long °	Sun Alt °	Sun Azm °	Path Width km	Central Line Dur	USA Geographic Regions	
2411 Jun 21	20:37:42	833	6326	153	A	0.3537	0.9467	44.2N	128.0W	69	173	210	05m46s	L48	--
2415 Apr 10	02:59:34	847	6373	155	T	0.5866	1.0437	38.9N	118.5E	54	144	178	03m16s	---	AK
2417 Aug 13	02:28:05	855	6402	142	T	0.6190	1.0723	48.3N	170.3E	52	216	297	04m55s	---	HI
2421 May 31	18:32:58	868	6449	144	A	0.3452	0.9750	42.4N	96.1W	70	177	95	02m32s	L48 + HI	
2422 Nov 14	17:27:39	874	6467	159	A	0.5657	0.9386	15.8N	76.9W	55	189	275	08m01s	L48	--
2424 Mar 31	02:10:09	879	6484	136	T	0.7652	1.0254	51.3N	130.8E	40	152	133	01m55s	---	AK
2440 Nov 25	01:18:38	940	6690	159	A	0.5445	0.9347	12.2N	163.8E	57	185	290	08m51s	---	HI
2444 Sep 12	17:35:34	954	6737	161	T	0.5548	1.0688	35.7N	69.0W	56	202	268	05m24s	L48	--
2449 Nov 15	20:23:55	973	6801	140	An	0.9810	0.9186	54.9N	90.2W	10	214	-	07m35s	L48	--
2451 May 01	18:53:36	979	6819	155	T	0.4957	1.0459	42.1N	115.4W	60	149	175	03m28s	L48	--
2460 Apr 21	18:09:48	1014	6930	136	T	0.8503	1.0237	66.9N	120.9W	31	142	154	01m34s	---	AK
2466 Jul 12	17:50:50	1038	7007	163	A	0.8460	0.9676	79.5N	66.9W	32	196	221	02m18s	---	AK
2469 May 12	02:39:06	1049	7042	155	T	0.4416	1.0466	42.6N	131.7E	64	153	172	03m36s	---	AK
2471 Sep 15	01:29:10	1058	7071	142	T	0.8109	1.0585	48.0N	163.3W	36	226	323	03m54s	---	AK
2472 Sep 03	16:12:53	1062	7083	152	A	0.0858	1.0255	11.3N	56.5W	85	208	87	02m19s	L48	--
2475 Jul 03	15:05:21	1074	7118	144	A	0.5775	0.9858	57.3N	28.9W	54	199	62	01m11s	L48	--
2483 Aug 03	22:21:16	1106	7218	153	A	0.0029	0.9479	17.4N	148.9W	90	186	192	06m50s	---	HI
2484 Jul 23	00:34:33	1110	7230	163	A	0.7618	0.9720	68.9N	167.2W	40	198	156	02m10s	L48 + AK	
2493 Jul 13	21:56:35	1147	7341	144	A	0.6562	0.9882	60.3N	122.5W	49	209	55	00m56s	L48 + AK	
2497 May 02	19:01:51	1163	7388	146	A	0.2341	0.9728	29.2N	104.4W	76	167	100	02m59s	L48	--
2498 Oct 15	17:34:43	1169	7406	161	T	0.4130	1.0597	14.6N	75.6W	66	196	215	05m21s	L48	--
2500 Mar 01	14:14:45	1175	7423	138	H	0.9038	1.0026	53.9N	51.9W	25	151	21	00m12s	L48	--
2505 Jun 03	17:48:01	1197	7488	155	T	0.3164	1.0464	40.5N	87.5W	71	165	163	03m50s	L48	--
2509 Mar 22	00:20:45	1213	7535	157	H	0.4675	1.0023	24.8N	167.1E	62	148	9	00m12s	---	AK
2518 Mar 12	22:37:00	1252	7646	138	T	0.9200	1.0071	59.1N	175.5E	23	144	63	00m31s	---	AK
2520 Aug 14	14:11:40	1262	7676	163	A	0.5983	0.9784	49.7N	14.1W	53	200	96	01m57s	L48	--
2525 Oct 18	01:23:53	1285	7740	142	T	0.9559	1.0396	52.7N	153.7W	17	227	450	02m39s	---	AK
2526 Oct 07	14:54:19	1289	7752	152	H	0.2558	1.0070	7.5N	34.2W	75	208	25	00m40s	L48	--
2537 Mar 12	21:49:05	1336	7881	148	T	0.2254	1.0542	9.5N	143.4W	77	162	184	04m53s	L48	--
2538 Aug 25	21:08:12	1342	7899	163	A	0.5217	0.9806	40.7N	120.3W	58	200	81	01m52s	L48	--
2551 Jun 05	16:10:38	1401	8057	146	A	0.4412	0.9699	49.0N	57.3W	64	179	121	02m55s	L48	--
2559 Jan 09	19:24:27	1436	8151	150	A	0.3841	0.9308	0.6N	104.5W	67	178	280	09m43s	---	HI
2563 Apr 24	00:08:29	1456	8204	157	T	0.3473	1.0214	31.1N	173.6E	70	151	77	01m49s	L48	--
2565 Aug 27	01:53:54	1467	8233	144	A	0.9527	0.9900	63.1N	130.5W	17	258	117	00m39s	L48	--
2566 Aug 16	14:22:23	1472	8245	154	T	0.1582	1.0569	21.8N	24.0W	81	206	190	04m47s	L48	--
2569 Jun 15	23:00:06	1485	8280	146	A	0.5197	0.9680	54.8N	154.9W	58	186	135	02m56s	L48 + AK	
2577 Jul 16	23:05:21	1524	8380	155	A	0.0230	1.0382	22.5N	158.0W	89	184	128	03m47s	---	HI
2580 Nov 08	14:34:16	1540	8421	152	A	0.3704	0.9883	3.4N	28.3W	68	201	44	01m15s	L48	--
2591 Apr 14	22:49:04	1591	8550	148	T	0.3190	1.0637	27.7N	161.2W	71	163	220	05m19s	L48	--
2613 Feb 11	19:51:41	1700	8820	150	A	0.4077	0.9382	9.6N	113.0W	66	167	250	08m00s	L48	--
2614 Jul 28	22:14:46	1707	8838	165	A	0.6680	0.9721	60.2N	133.2W	48	196	135	02m21s	L48 + AK	
2617 May 26	23:01:02	1722	8873	157	T	0.1740	1.0394	31.0N	161.7W	80	163	134	03m30s	L48	--
2618 May 16	14:44:45	1727	8885	167	T	0.8918	1.0612	70.9N	98.3W	27	109	447	03m24s	L48	--
2624 Jul 07	20:02:08	1759	8961	156	Am	0.0151	0.9487	23.4N	111.4W	89	191	188	06m24s	---	HI
2634 Dec 12	15:11:10	1813	9090	152	A	0.4304	0.9723	2.2N	37.8W	64	189	110	03m19s	L48	--
2638 Sep 29	21:06:35	1833	9137	154	T	0.4008	1.0554	18.1N	119.6W	66	210	198	04m31s	---	AK
2641 Jul 30	01:35:53	1848	9172	146	A	0.8602	0.9545	68.1N	141.2W	30	242	326	03m20s	L48 + AK	
2645 May 17	22:43:15	1868	9219	148	T	0.4686	1.0707	47.4N	158.9W	62	170	261	05m16s	L48	--
2654 Jun 07	06:09:48	1917	9331	167	T	0.7664	1.0703	70.6N	69.0E	40	148	358	04m12s	---	AK
2657 Apr 06	00:01:43	1933	9366	159	A	0.3349	0.9305	23.6N	178.3E	70	149	274	07m38s	L48	--
2661 Jan 22	23:02:20	1953	9413	161	T	0.2739	1.0324	4.4S	160.0W	74	162	113	03m04s	L48	--
2667 Mar 16	19:47:37	1987	9489	150	A	0.4614	0.9506	24.6N	115.4W	62	161	203	05m36s	L48	--
2668 Aug 29	18:33:46	1995	9507	165	A	0.4359	0.9627	33.7N	81.4W	64	199	149	03m59s	L48	--
2672 Jun 17	13:46:14	2016	9554	167	T	0.6986	1.0735	67.3N	29.8W	45	164	335	04m36s	L48	--
2678 Aug 09	15:23:53	2051	9630	156	A	0.2783	0.9492	30.4N	33.4W	74	206	194	05m30s	L48	--
2680 Jan 23	19:32:54	2059	9648	171	An	0.9969	0.9636	62.0N	145.2W	2	140	-	02m46s	---	AK
2681 Jun 08	14:07:29	2067	9665	148	T	0.5954	1.0724	59.7N	22.3W	53	181	294	04m54s	L48	--
2683 Nov 10	22:07:52	2080	9695	173	T	0.8265	1.0331	37.9N	136.5W	34	193	198	02m49s	L48 + AK	
2690 Jun 28	21:18:05	2118	9777	167	T	0.6272	1.0759	62.2N	132.5W	51	176	317	05m00s	---	AK
2692 Oct 31	21:27:37	2131	9806	154	T	0.5212	1.0503	14.4N	124.5W	58	204	194	04m23s	---	AK

L48 = Lower 48 States   *AK = Alaska*   *HI = Hawaii*

33

Calendar Date	TD of Greatest Eclipse	ΔT s	Luna Num	Saros Num	Ecl Type	Gamma	Ecl Mag	Lat °	Long °	Sun Alt °	Sun Azm °	Path Width km	Central Line Dur	USA Geographic Regions	
2694 Apr 16	15:01:07	2140	9824	169	A	0.9488	0.9422	63.2N	107.0W	18	103	679	04m05s	L48	--
2699 Jun 19	21:42:29	2170	9888	148	T	0.6645	1.0720	64.9N	127.7W	48	191	314	04m38s	L48	+ AK
2707 Jan 26	00:42:03	2214	9982	152	A	0.4647	0.9587	8.5N	177.6E	62	172	169	05m08s	----	HI
2711 May 09	21:21:38	2239	10035	159	A	0.1700	0.9385	26.5N	135.8W	80	157	231	07m05s	L48	--
2719 Dec 03	15:07:58	2289	10141	173	T	0.7836	1.0331	29.9N	38.0W	38	183	180	03m01s	L48	--
2721 Apr 18	18:47:23	2298	10158	150	A	0.5666	0.9657	44.1N	104.2W	55	159	150	03m17s	L48	+ HI
2722 Oct 02	15:28:03	2306	10176	165	A	0.2383	0.9501	9.7N	40.8W	76	197	188	06m12s	L48	--
2728 Nov 23	14:20:43	2343	10252	154	T	0.5701	1.0468	13.1N	18.9W	55	197	188	04m17s	L48	--
2734 Feb 25	20:45:23	2375	10317	171	A	0.9773	0.9615	56.1N	171.5W	11	126	673	02m55s	----	AK
2737 Dec 13	23:47:40	2398	10364	173	T	0.7697	1.0332	27.4N	170.4W	39	178	176	03m04s	L48	--
2739 Apr 30	02:11:53	2406	10381	150	A	0.6158	0.9708	51.4N	143.9E	52	160	133	02m37s	----	AK
2741 Oct 01	22:44:39	2421	10411	175	A	0.9162	0.9303	58.5N	122.0W	23	215	651	06m14s	L48	+ AK
2744 Jul 31	19:48:22	2438	10446	167	T	0.4081	1.0778	41.8N	99.7W	66	192	276	05m59s	L48	--
2748 May 19	12:36:21	2462	10493	169	A	0.7708	0.9624	65.1N	34.3W	39	136	213	02m53s	L48	--
2750 Sep 22	18:14:41	2476	10522	156	A	0.5818	0.9445	30.9N	65.8W	54	214	246	05m40s	L48	+ AK
2766 May 30	19:35:59	2574	10716	169	A	0.6995	0.9686	63.7N	126.4W	45	149	158	02m29s	----	AK
2768 Oct 03	01:16:24	2589	10745	156	A	0.6427	0.9428	30.7N	169.9W	50	214	269	05m54s	----	HI
2770 Mar 19	13:02:29	2598	10763	171	A	0.9422	0.9624	55.6N	56.6W	19	122	401	02m48s	L48	--
2771 Aug 03	03:38:31	2607	10780	148	T	0.9591	1.0590	69.6N	130.5W	16	277	704	03m05s	L48	+ AK
2772 Jul 22	18:22:11	2613	10792	158	T	0.2260	1.0272	32.4N	77.8W	77	200	95	02m27s	L48	--
2774 Jan 04	17:20:28	2622	10810	173	T	0.7520	1.0342	25.4N	77.9W	41	168	174	03m07s	L48	--
2775 May 21	16:45:17	2631	10827	150	A	0.7292	0.9804	66.7N	72.7W	43	165	102	01m31s	L48	--
2782 Dec 26	16:26:44	2680	10921	154	T	0.6071	1.0435	14.1N	53.5W	53	184	183	04m10s	L48	--
2784 Jun 10	02:32:13	2689	10939	169	A	0.6244	0.9744	60.8N	140.6E	51	161	118	02m07s	L48	+ AK
2788 Mar 29	20:59:23	2714	10986	171	A	0.9158	0.9633	56.3N	174.6W	23	121	325	02m43s	----	AK
2792 Jan 16	02:09:54	2738	11033	173	T	0.7453	1.0353	25.6N	147.7E	42	164	177	03m09s	----	AK
2795 Nov 03	20:29:00	2763	11080	175	A	0.7682	0.9285	34.2N	108.5W	40	194	416	08m26s	L48	+ AK
2797 Mar 20	17:29:16	2772	11097	152	A	0.5455	0.9534	31.6N	80.2W	57	159	202	05m00s	L48	--
2798 Sep 02	18:30:48	2782	11115	167	T	0.2008	1.0719	18.7N	82.5W	78	197	238	06m14s	L48	--
2805 Apr 20	18:41:05	2826	11197	161	T	0.1129	1.0313	17.7N	91.7W	83	154	106	02m46s	L48	--
2811 Jun 12	06:58:42	2866	11273	150	A	0.8624	0.9880	82.8N	101.4E	30	195	84	00m47s	----	AK
2812 May 31	18:39:54	2873	11285	160	Tm	0.0695	1.0493	26.2N	88.3W	86	178	164	04m36s	L48	--
2820 Jul 01	16:16:59	2927	11385	169	A	0.4651	0.9847	50.8N	51.3W	62	179	61	01m24s	L48	--
2822 Nov 04	23:19:08	2943	11414	156	A	0.7818	0.9376	31.9N	140.4W	38	207	362	06m49s	L48	+ AK
2828 Feb 06	19:50:40	2978	11479	173	T	0.7326	1.0388	28.0N	121.3W	43	155	188	03m15s	L48	+ HI
2829 Jun 22	14:01:21	2988	11496	150	A	0.9336	0.9904	83.4N	97.0E	21	296	97	00m35s	----	AK
2835 Sep 14	00:33:14	3030	11573	177	T	0.8440	1.0151	56.9N	148.8W	32	213	96	01m07s	L48	+ AK
2837 Jan 27	19:01:12	3040	11590	154	T	0.6224	1.0438	20.0N	96.0W	51	171	187	04m02s	L48	--
2838 Jul 12	23:08:11	3050	11608	169	A	0.3827	0.9891	44.3N	150.7W	67	184	42	01m03s	L48	--
2842 May 01	19:50:31	3076	11655	171	A	0.7925	0.9664	59.9N	144.2W	37	131	197	02m33s	----	AK
2844 Sep 03	23:05:33	3092	11684	158	H	0.5279	1.0063	34.5N	136.0W	58	214	25	00m32s	----	AK
2849 Dec 05	19:34:15	3129	11749	175	A	0.6813	0.9266	20.7N	102.4W	47	181	377	09m51s	L48	--
2851 Apr 22	16:22:27	3138	11766	152	A	0.6645	0.9529	52.4N	68.3W	48	157	230	04m20s	L48	--
2863 Mar 10	22:51:03	3222	11913	163	H	0.0298	1.0147	2.3S	147.9W	88	151	50	01m21s	L48	--
2866 Jul 03	17:03:11	3245	11954	160	T	0.2786	1.0610	38.7N	56.7W	74	193	209	04m59s	L48	--
2869 May 02	23:40:18	3265	11989	152	A	0.7193	0.9525	60.4N	179.1W	44	157	250	04m05s	----	AK
2876 Dec 06	22:44:49	3320	12083	156	A	0.8571	0.9349	35.5N	137.4W	31	193	468	07m22s	L48	+ AK
2882 Mar 10	21:58:27	3358	12148	173	T	0.6933	1.0475	34.4N	157.1W	46	145	215	03m34s	----	AK
2889 Oct 15	23:21:03	3413	12242	177	H	0.6917	1.0004	33.0N	146.5W	46	199	2	00m02s	L48	--
2891 Mar 01	21:22:00	3423	12259	154	T	0.6502	1.0477	31.7N	136.1W	49	161	208	03m58s	L48	--
2896 Jun 02	17:19:09	3461	12324	171	A	0.6070	0.9683	58.5N	79.9W	52	158	144	02m42s	L48	--
2898 Oct 06	21:13:36	3479	12353	158	A	0.7154	0.9864	34.1N	102.7W	44	215	67	01m13s	L48	+ AK
2904 Jan 08	19:40:26	3517	12418	175	A	0.6385	0.9281	16.7N	108.0W	50	167	348	09m24s	L48	--
2905 May 25	13:49:00	3528	12435	152	A	0.8483	0.9505	78.6N	33.4W	32	157	346	03m39s	L48	--
2914 Jun 15	00:12:34	3595	12547	171	A	0.5336	0.9681	55.2N	175.1W	57	167	136	02m52s	L48	+ AK
2920 Aug 05	15:19:05	3641	12623	160	T	0.4992	1.0660	43.8N	17.4W	60	210	248	04m48s	L48	--
2922 Jan 19	03:49:58	3652	12641	175	A	0.6294	0.9296	17.2N	128.5E	51	163	335	08m53s	----	AK
2924 May 24	21:17:48	3670	12670	162	A	0.1806	0.9426	31.5N	125.5W	79	175	216	07m02s	L48	+ HI
2925 Nov 07	15:05:09	3681	12688	177	A	0.6201	0.9895	21.4N	27.7W	52	192	47	01m08s	L48	--
2927 Mar 24	14:31:58	3691	12705	154	T	0.6887	1.0514	42.6N	37.7W	46	156	233	03m54s	L48	--

L48 = Lower 48 States   AK = Alaska   HI = Hawaii

Calendar Date	TD of Greatest Eclipse	ΔT s	Luna Num	Saros Num	Ecl Type	Gamma	Ecl Mag	Lat °	Long °	Sun Alt °	Sun Azm °	Path Width km	Central Line Dur	USA Geographic Regions	
2929 Aug 25	22:47:38	3710	12735	179	T	0.7412	1.0515	55.6N	125.2W	42	207	254	03m37s	L48 +	AK
2935 Oct 18	16:07:45	3756	12811	168	A	0.0926	0.9401	5.0S	47.5W	85	207	223	06m59s	L48	--
2936 Apr 12	23:13:32	3760	12817	173	T	0.6096	1.0584	41.6N	174.4W	52	143	240	04m08s	---	AK
2938 Aug 16	22:48:58	3778	12846	160	T	0.5698	1.0660	43.9N	125.5W	55	215	261	04m42s	L48 +	AK
2943 Nov 18	23:06:13	3819	12911	177	A	0.5930	0.9840	17.0N	149.7W	54	188	70	01m48s	L48 +	HI
2945 Apr 03	22:56:45	3829	12928	154	T	0.7165	1.0532	49.0N	166.1W	44	154	251	03m50s	---	AK
2952 Nov 08	20:16:35	3888	13022	158	A	0.8490	0.9660	37.2N	88.6W	32	207	227	03m18s	L48	--
2958 Feb 09	20:08:45	3929	13087	175	A	0.6087	0.9347	20.0N	118.2W	52	155	300	07m33s	L48	--
2967 Jan 31	15:44:44	4000	13198	156	A	0.8962	0.9413	45.7N	49.9W	26	166	491	05m54s	L48	--
2968 Jul 16	20:21:11	4011	13216	171	A	0.2908	0.9651	37.9N	104.7W	73	186	132	03m48s	L48	--
2978 Jun 26	17:12:30	4090	13339	162	A	0.4167	0.9493	47.7N	54.6W	65	191	205	05m12s	L48	--
2979 Dec 10	15:27:20	4102	13357	177	A	0.5562	0.9740	11.1N	37.3W	56	179	112	03m07s	L48	--
2981 Apr 25	15:22:34	4113	13374	154	T	0.7918	1.0560	63.4N	58.3W	37	150	303	03m36s	L48	--
2985 Feb 10	23:59:57	4143	13421	156	A	0.9029	0.9444	49.0N	178.0W	25	160	477	05m18s	---	AK
2990 May 15	23:21:58	4186	13486	173	T	0.4709	1.0689	45.4N	166.4W	62	154	254	04m58s	L48 +	AK
2992 Sep 17	21:42:03	4205	13515	160	T	0.7637	1.0617	43.5N	98.7W	40	221	307	04m16s	L48	--
2993 Sep 07	14:40:07	4213	13527	170	T	0.0388	1.0673	7.4N	21.6W	88	208	220	05m33s	L48	--
2996 Jul 06	23:43:58	4236	13562	162	A	0.5014	0.9508	51.6N	146.3W	60	199	208	04m44s	L48 +	AK
2997 Dec 20	23:45:10	4247	13580	177	A	0.5448	0.9697	9.6N	162.5W	57	175	130	03m40s	L48	--

L48 = Lower 48 States   AK = Alaska   HI = Hawaii

Appendix B

Global Maps of Central Solar Eclipses in the USA

Key to Solar Eclipse Maps

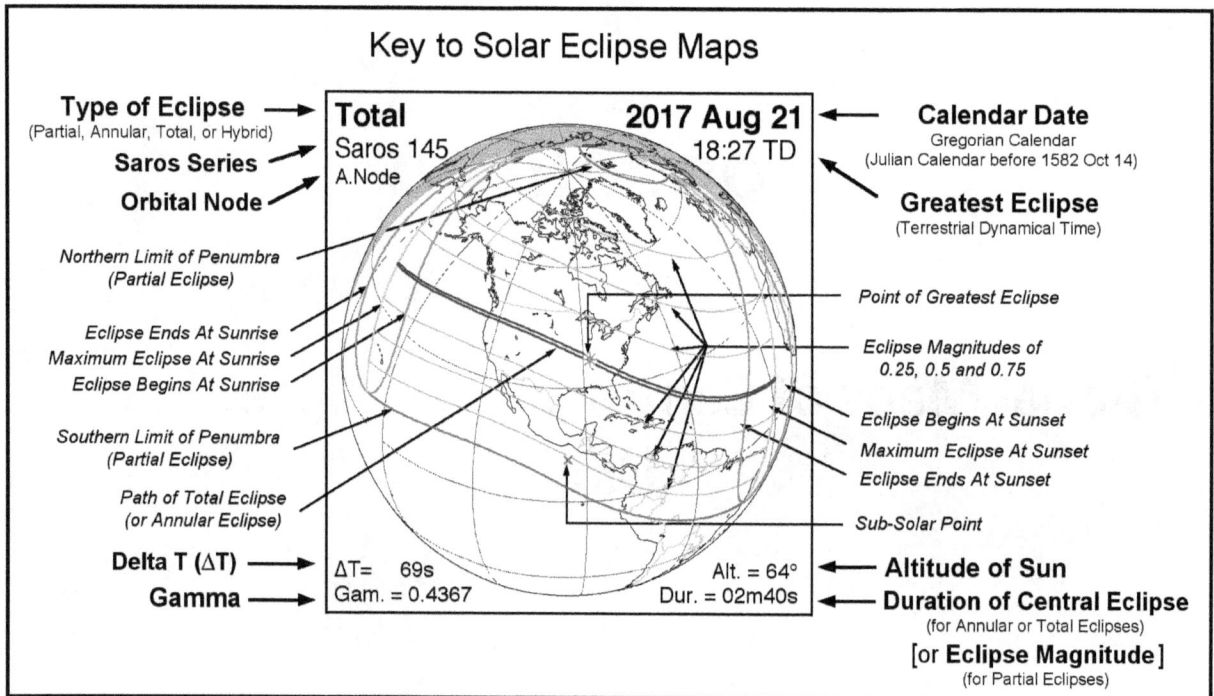

Key to Solar Eclipse Maps

Type of Eclipse →
(Partial, Annular, Total, or Hybrid)

Saros Series →

Orbital Node →

Northern Limit of Penumbra
(Partial Eclipse)

Eclipse Ends At Sunrise
Maximum Eclipse At Sunrise
Eclipse Begins At Sunrise

Southern Limit of Penumbra
(Partial Eclipse)

Path of Total Eclipse
(or Annular Eclipse)

Delta T (ΔT) →

Gamma →

Total
Saros 145
A.Node

2017 Aug 21
18:27 TD

← **Calendar Date**
Gregorian Calendar
(Julian Calendar before 1582 Oct 14)

Greatest Eclipse
(Terrestrial Dynamical Time)

Point of Greatest Eclipse

Eclipse Magnitudes of
0.25, 0.5 and 0.75

Eclipse Begins At Sunset
Maximum Eclipse At Sunset
Eclipse Ends At Sunset

Sub-Solar Point

ΔT=   69s
Gam. = 0.4367

Alt. = 64°
Dur. = 02m40s

← **Altitude of Sun**
← **Duration of Central Eclipse**
(for Annular or Total Eclipses)
[or **Eclipse Magnitude**]
(for Partial Eclipses)

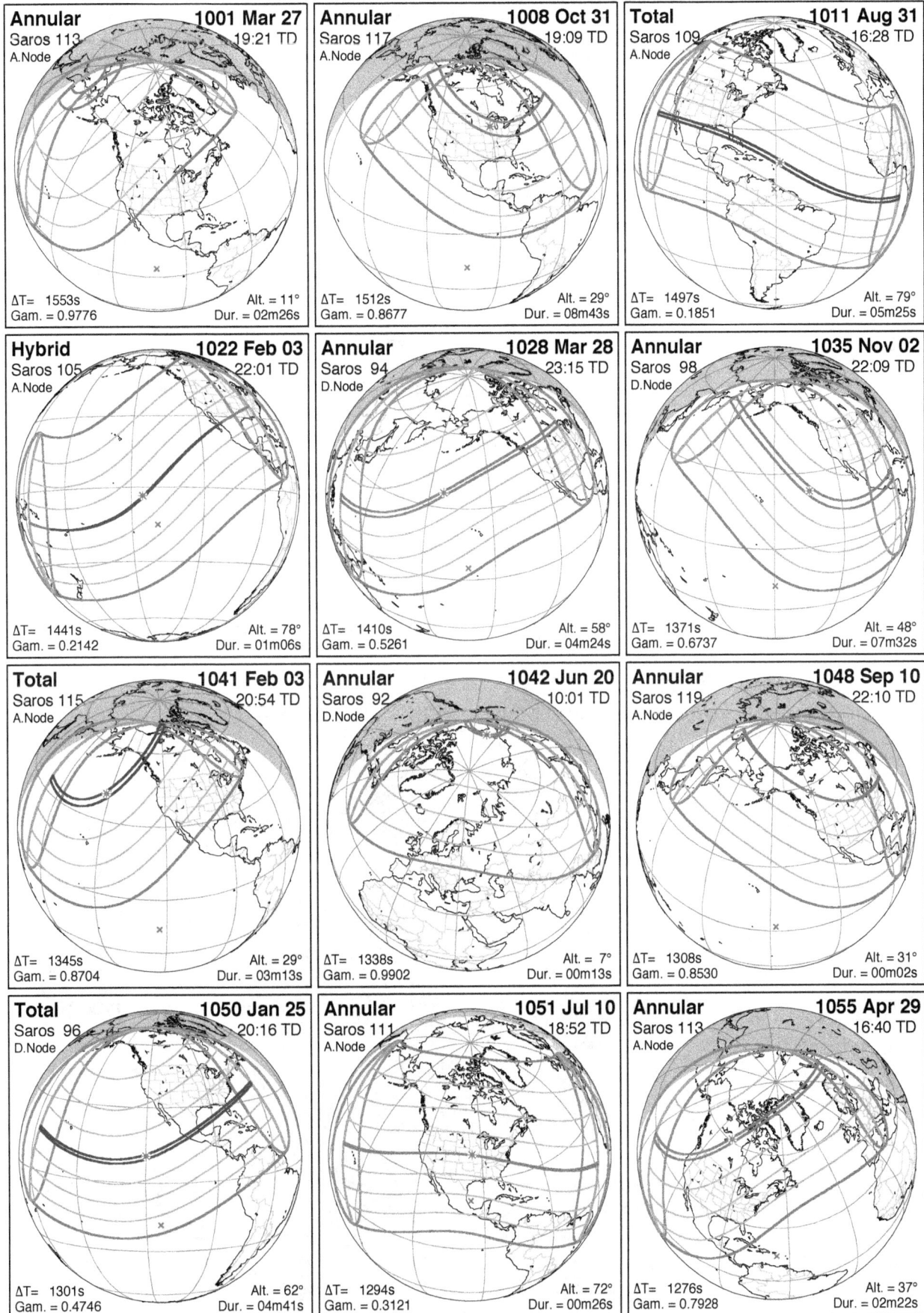

Annular **1001 Mar 27**	
Saros 113 19:21 TD	
A.Node	
ΔT= 1553s Alt. = 11°	
Gam. = 0.9776 Dur. = 02m26s	

Annular   **1001 Mar 27** — Saros 113, A.Node, 19:21 TD, ΔT= 1553s, Gam. = 0.9776, Alt. = 11°, Dur. = 02m26s

Annular   **1008 Oct 31** — Saros 117, A.Node, 19:09 TD, ΔT= 1512s, Gam. = 0.8677, Alt. = 29°, Dur. = 08m43s

Total   **1011 Aug 31** — Saros 109, A.Node, 16:28 TD, ΔT= 1497s, Gam. = 0.1851, Alt. = 79°, Dur. = 05m25s

Hybrid   **1022 Feb 03** — Saros 105, A.Node, 22:01 TD, ΔT= 1441s, Gam. = 0.2142, Alt. = 78°, Dur. = 01m06s

Annular   **1028 Mar 28** — Saros 94, D.Node, 23:15 TD, ΔT= 1410s, Gam. = 0.5261, Alt. = 58°, Dur. = 04m24s

Annular   **1035 Nov 02** — Saros 98, D.Node, 22:09 TD, ΔT= 1371s, Gam. = 0.6737, Alt. = 48°, Dur. = 07m32s

Total   **1041 Feb 03** — Saros 115, A.Node, 20:54 TD, ΔT= 1345s, Gam. = 0.8704, Alt. = 29°, Dur. = 03m13s

Annular   **1042 Jun 20** — Saros 92, D.Node, 10:01 TD, ΔT= 1338s, Gam. = 0.9902, Alt. = 7°, Dur. = 00m13s

Annular   **1048 Sep 10** — Saros 119, A.Node, 22:10 TD, ΔT= 1308s, Gam. = 0.8530, Alt. = 31°, Dur. = 00m02s

Total   **1050 Jan 25** — Saros 96, D.Node, 20:16 TD, ΔT= 1301s, Gam. = 0.4746, Alt. = 62°, Dur. = 04m41s

Annular   **1051 Jul 10** — Saros 111, A.Node, 18:52 TD, ΔT= 1294s, Gam. = 0.3121, Alt. = 72°, Dur. = 00m26s

Annular   **1055 Apr 29** — Saros 113, A.Node, 16:40 TD, ΔT= 1276s, Gam. = 0.7928, Alt. = 37°, Dur. = 02m22s

Plate 001

39

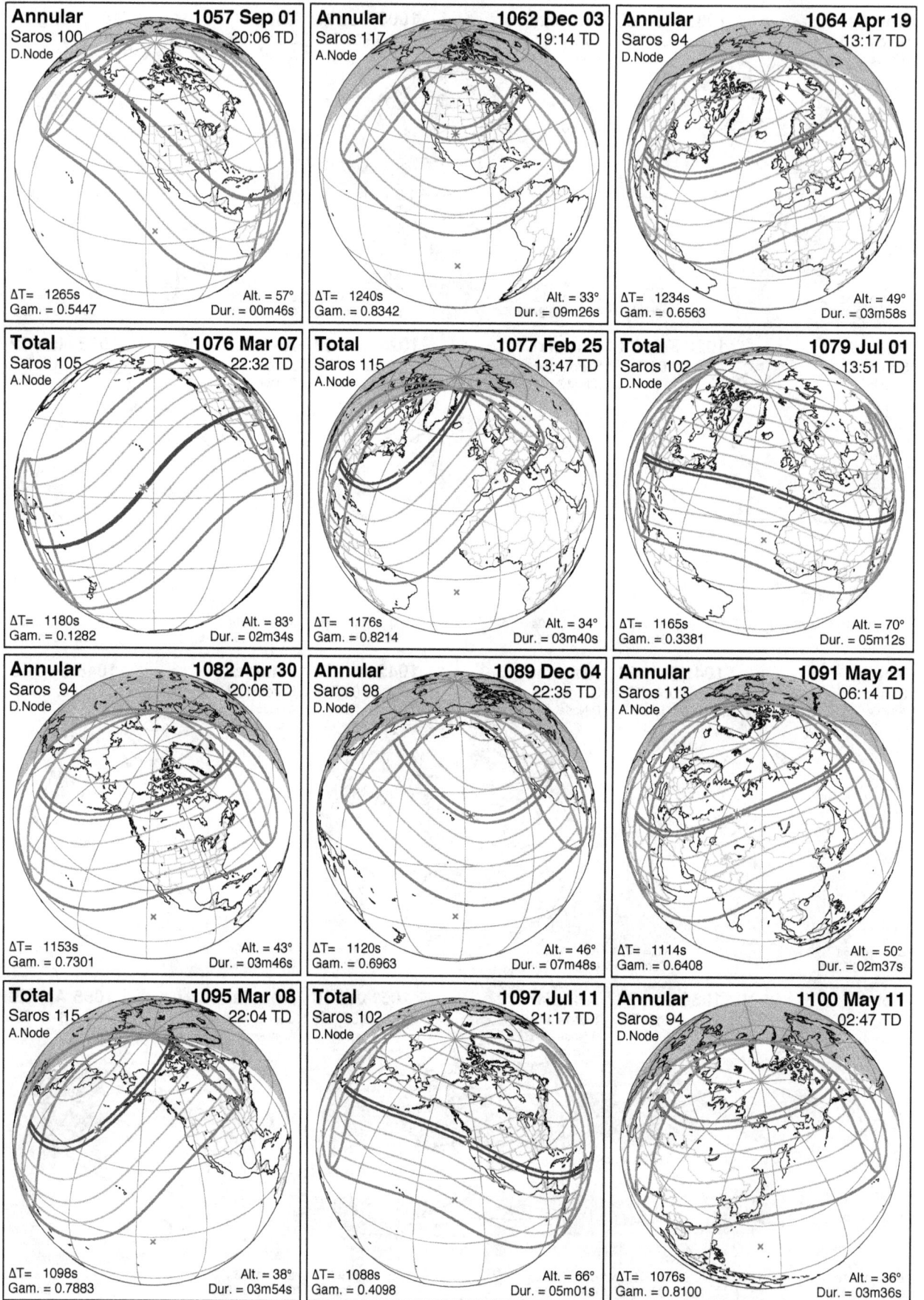

Annular **1057 Sep 01**	
Saros 100 20:06 TD	
D.Node	
ΔT= 1265s Alt. = 57°	
Gam. = 0.5447 Dur. = 00m46s	

Annular   **1057 Sep 01**
Saros 100   20:06 TD
D.Node
ΔT= 1265s   Alt. = 57°
Gam. = 0.5447   Dur. = 00m46s

Annular   **1062 Dec 03**
Saros 117   19:14 TD
A.Node
ΔT= 1240s   Alt. = 33°
Gam. = 0.8342   Dur. = 09m26s

Annular   **1064 Apr 19**
Saros 94   13:17 TD
D.Node
ΔT= 1234s   Alt. = 49°
Gam. = 0.6563   Dur. = 03m58s

Total   **1076 Mar 07**
Saros 105   22:32 TD
A.Node
ΔT= 1180s   Alt. = 83°
Gam. = 0.1282   Dur. = 02m34s

Total   **1077 Feb 25**
Saros 115   13:47 TD
A.Node
ΔT= 1176s   Alt. = 34°
Gam. = 0.8214   Dur. = 03m40s

Total   **1079 Jul 01**
Saros 102   13:51 TD
D.Node
ΔT= 1165s   Alt. = 70°
Gam. = 0.3381   Dur. = 05m12s

Annular   **1082 Apr 30**
Saros 94   20:06 TD
D.Node
ΔT= 1153s   Alt. = 43°
Gam. = 0.7301   Dur. = 03m46s

Annular   **1089 Dec 04**
Saros 98   22:35 TD
D.Node
ΔT= 1120s   Alt. = 46°
Gam. = 0.6963   Dur. = 07m48s

Annular   **1091 May 21**
Saros 113   06:14 TD
A.Node
ΔT= 1114s   Alt. = 50°
Gam. = 0.6408   Dur. = 02m37s

Total   **1095 Mar 08**
Saros 115   22:04 TD
A.Node
ΔT= 1098s   Alt. = 38°
Gam. = 0.7883   Dur. = 03m54s

Total   **1097 Jul 11**
Saros 102   21:17 TD
D.Node
ΔT= 1088s   Alt. = 66°
Gam. = 0.4098   Dur. = 05m01s

Annular   **1100 May 11**
Saros 94   02:47 TD
D.Node
ΔT= 1076s   Alt. = 36°
Gam. = 0.8100   Dur. = 03m36s

Plate 002

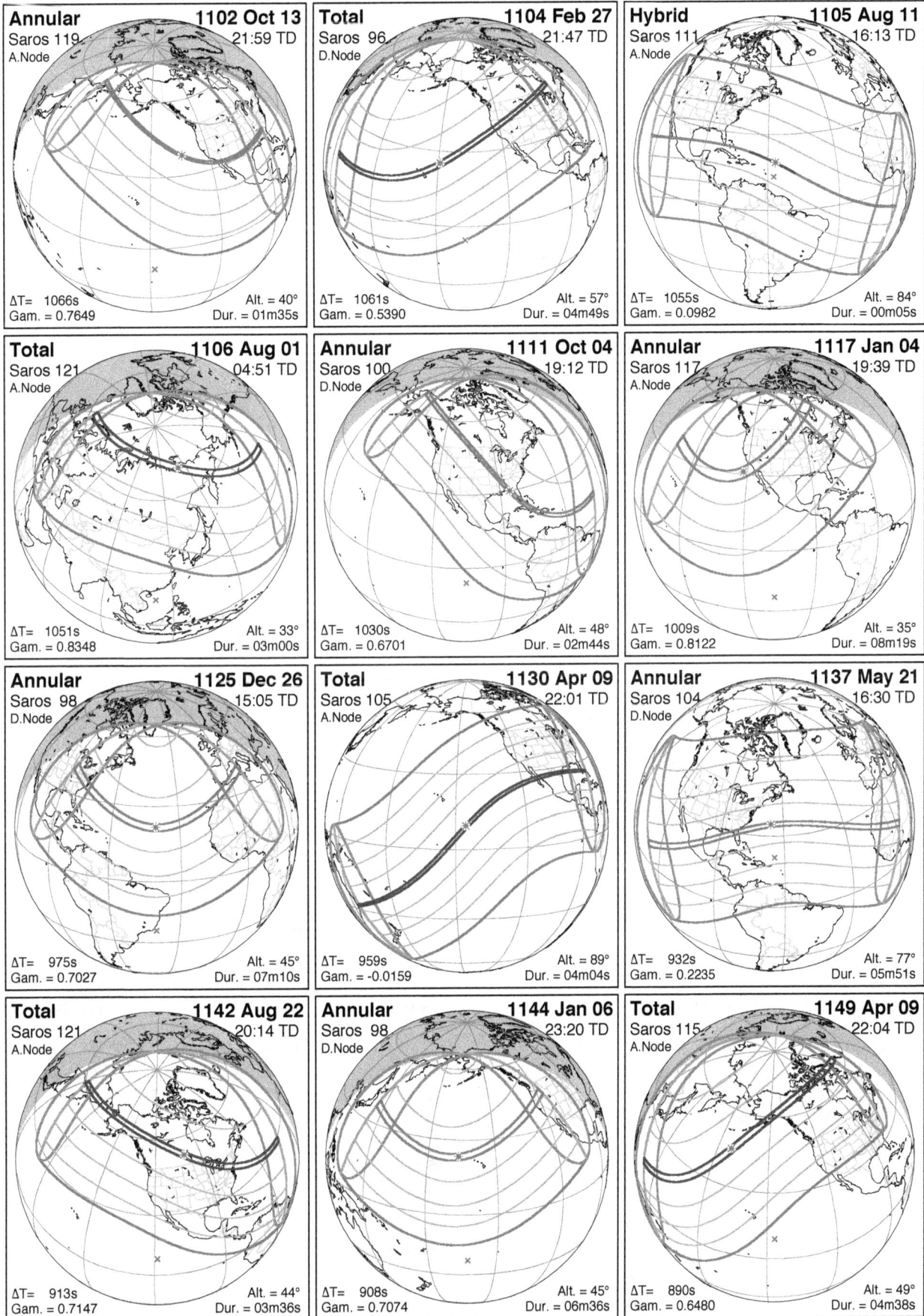

Annular	1102 Oct 13
Saros 119	21:59 TD
A.Node	
ΔT= 1066s	Alt. = 40°
Gam. = 0.7649	Dur. = 01m35s

Total	1104 Feb 27
Saros 96	21:47 TD
D.Node	
ΔT= 1061s	Alt. = 57°
Gam. = 0.5390	Dur. = 04m49s

Hybrid	1105 Aug 11
Saros 111	16:13 TD
A.Node	
ΔT= 1055s	Alt. = 84°
Gam. = 0.0982	Dur. = 00m05s

Total	1106 Aug 01
Saros 121	04:51 TD
A.Node	
ΔT= 1051s	Alt. = 33°
Gam. = 0.8348	Dur. = 03m00s

Annular	1111 Oct 04
Saros 100	19:12 TD
D.Node	
ΔT= 1030s	Alt. = 48°
Gam. = 0.6701	Dur. = 02m44s

Annular	1117 Jan 04
Saros 117	19:39 TD
A.Node	
ΔT= 1009s	Alt. = 35°
Gam. = 0.8122	Dur. = 08m19s

Annular	1125 Dec 26
Saros 98	15:05 TD
D.Node	
ΔT= 975s	Alt. = 45°
Gam. = 0.7027	Dur. = 07m10s

Total	1130 Apr 09
Saros 105	22:01 TD
A.Node	
ΔT= 959s	Alt. = 89°
Gam. = -0.0159	Dur. = 04m04s

Annular	1137 May 21
Saros 104	16:30 TD
D.Node	
ΔT= 932s	Alt. = 77°
Gam. = 0.2235	Dur. = 05m51s

Total	1142 Aug 22
Saros 121	20:14 TD
A.Node	
ΔT= 913s	Alt. = 44°
Gam. = 0.7147	Dur. = 03m36s

Annular	1144 Jan 06
Saros 98	23:20 TD
D.Node	
ΔT= 908s	Alt. = 45°
Gam. = 0.7074	Dur. = 06m36s

Total	1149 Apr 09
Saros 115	22:04 TD
A.Node	
ΔT= 890s	Alt. = 49°
Gam. = 0.6480	Dur. = 04m38s

Plate 003

41

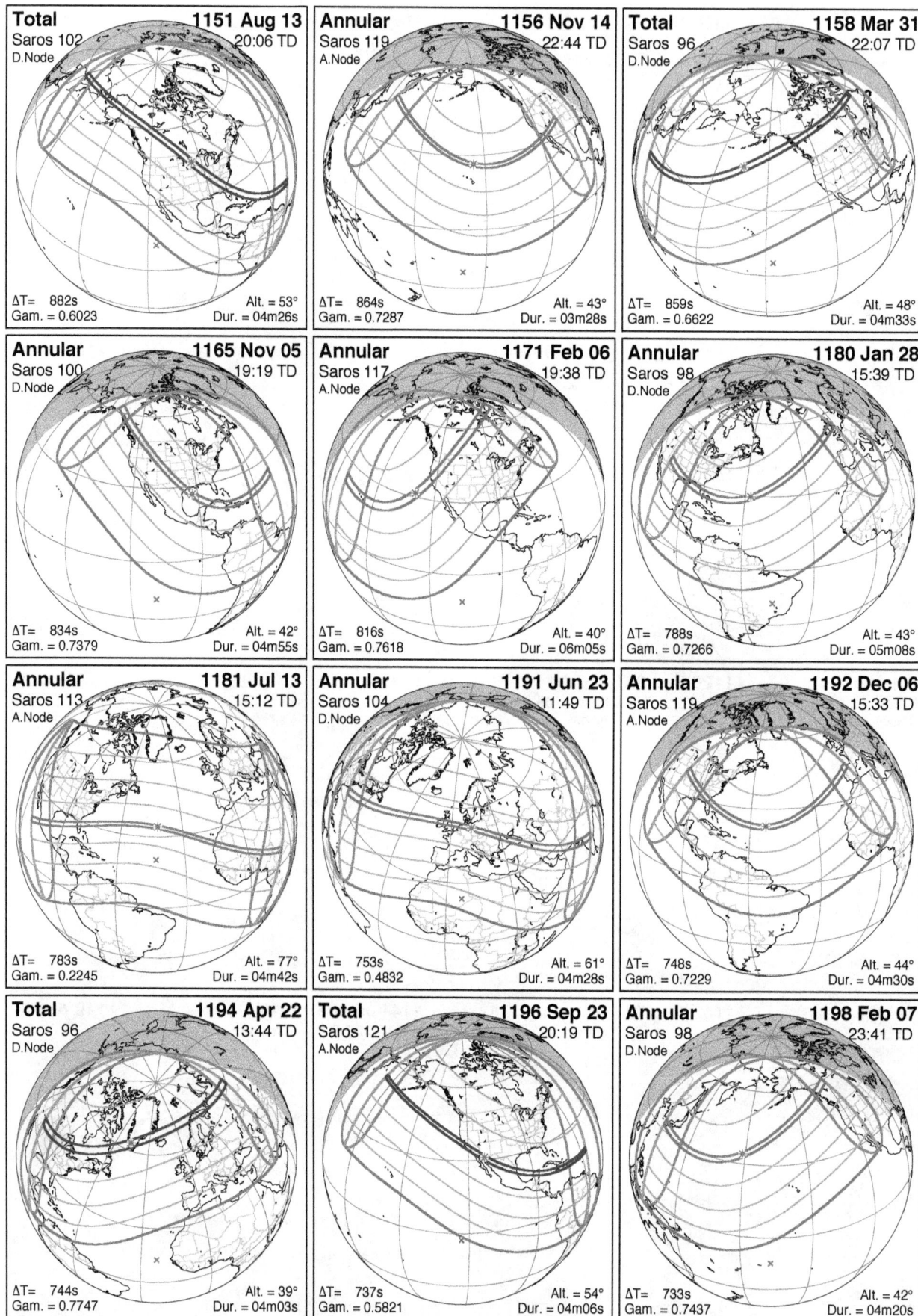

Total 1151 Aug 13 Saros 102 20:06 TD D.Node ΔT= 882s Alt. = 53° Gam. = 0.6023 Dur. = 04m26s	**Annular** 1156 Nov 14 Saros 119 22:44 TD A.Node ΔT= 864s Alt. = 43° Gam. = 0.7287 Dur. = 03m28s
Total 1158 Mar 31 Saros 96 22:07 TD D.Node ΔT= 859s Alt. = 48° Gam. = 0.6622 Dur. = 04m33s	

Annular 1165 Nov 05 Saros 100 19:19 TD D.Node ΔT= 834s Alt. = 42° Gam. = 0.7379 Dur. = 04m55s	**Annular** 1171 Feb 06 Saros 117 19:38 TD A.Node ΔT= 816s Alt. = 40° Gam. = 0.7618 Dur. = 06m05s	**Annular** 1180 Jan 28 Saros 98 15:39 TD D.Node ΔT= 788s Alt. = 43° Gam. = 0.7266 Dur. = 05m08s
Annular 1181 Jul 13 Saros 113 15:12 TD A.Node ΔT= 783s Alt. = 77° Gam. = 0.2245 Dur. = 04m42s	**Annular** 1191 Jun 23 Saros 104 11:49 TD D.Node ΔT= 753s Alt. = 61° Gam. = 0.4832 Dur. = 04m28s	**Annular** 1192 Dec 06 Saros 119 15:33 TD A.Node ΔT= 748s Alt. = 44° Gam. = 0.7229 Dur. = 04m30s
Total 1194 Apr 22 Saros 96 13:44 TD D.Node ΔT= 744s Alt. = 39° Gam. = 0.7747 Dur. = 04m03s	**Total** 1196 Sep 23 Saros 121 20:19 TD A.Node ΔT= 737s Alt. = 54° Gam. = 0.5821 Dur. = 04m06s	**Annular** 1198 Feb 07 Saros 98 23:41 TD D.Node ΔT= 733s Alt. = 42° Gam. = 0.7437 Dur. = 04m20s

Plate 004

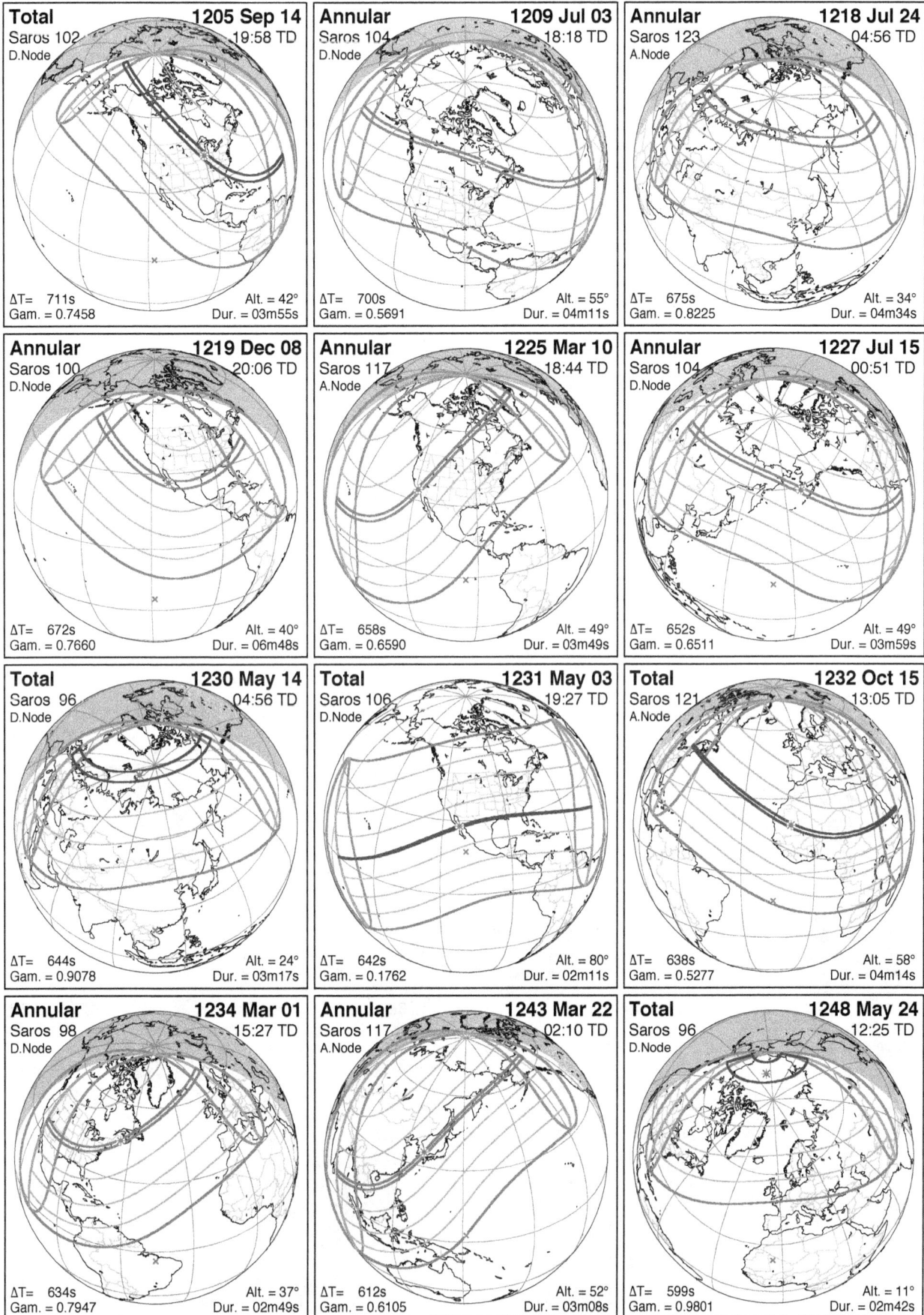

Total	1205 Sep 14
Saros 102	19:58 TD
D.Node	
ΔT= 711s	Alt. = 42°
Gam. = 0.7458	Dur. = 03m55s

Annular	1209 Jul 03
Saros 104	18:18 TD
D.Node	
ΔT= 700s	Alt. = 55°
Gam. = 0.5691	Dur. = 04m11s

Annular	1218 Jul 24
Saros 123	04:56 TD
A.Node	
ΔT= 675s	Alt. = 34°
Gam. = 0.8225	Dur. = 04m34s

Annular	1219 Dec 08
Saros 100	20:06 TD
D.Node	
ΔT= 672s	Alt. = 40°
Gam. = 0.7660	Dur. = 06m48s

Annular	1225 Mar 10
Saros 117	18:44 TD
A.Node	
ΔT= 658s	Alt. = 49°
Gam. = 0.6590	Dur. = 03m49s

Annular	1227 Jul 15
Saros 104	00:51 TD
D.Node	
ΔT= 652s	Alt. = 49°
Gam. = 0.6511	Dur. = 03m59s

Total	1230 May 14
Saros 96	04:56 TD
D.Node	
ΔT= 644s	Alt. = 24°
Gam. = 0.9078	Dur. = 03m17s

Total	1231 May 03
Saros 106	19:27 TD
D.Node	
ΔT= 642s	Alt. = 80°
Gam. = 0.1762	Dur. = 02m11s

Total	1232 Oct 15
Saros 121	13:05 TD
A.Node	
ΔT= 638s	Alt. = 58°
Gam. = 0.5277	Dur. = 04m14s

Annular	1234 Mar 01
Saros 98	15:27 TD
D.Node	
ΔT= 634s	Alt. = 37°
Gam. = 0.7947	Dur. = 02m49s

Annular	1243 Mar 22
Saros 117	02:10 TD
A.Node	
ΔT= 612s	Alt. = 52°
Gam. = 0.6105	Dur. = 03m08s

Total	1248 May 24
Saros 96	12:25 TD
D.Node	
ΔT= 599s	Alt. = 11°
Gam. = 0.9801	Dur. = 02m42s

Plate 005

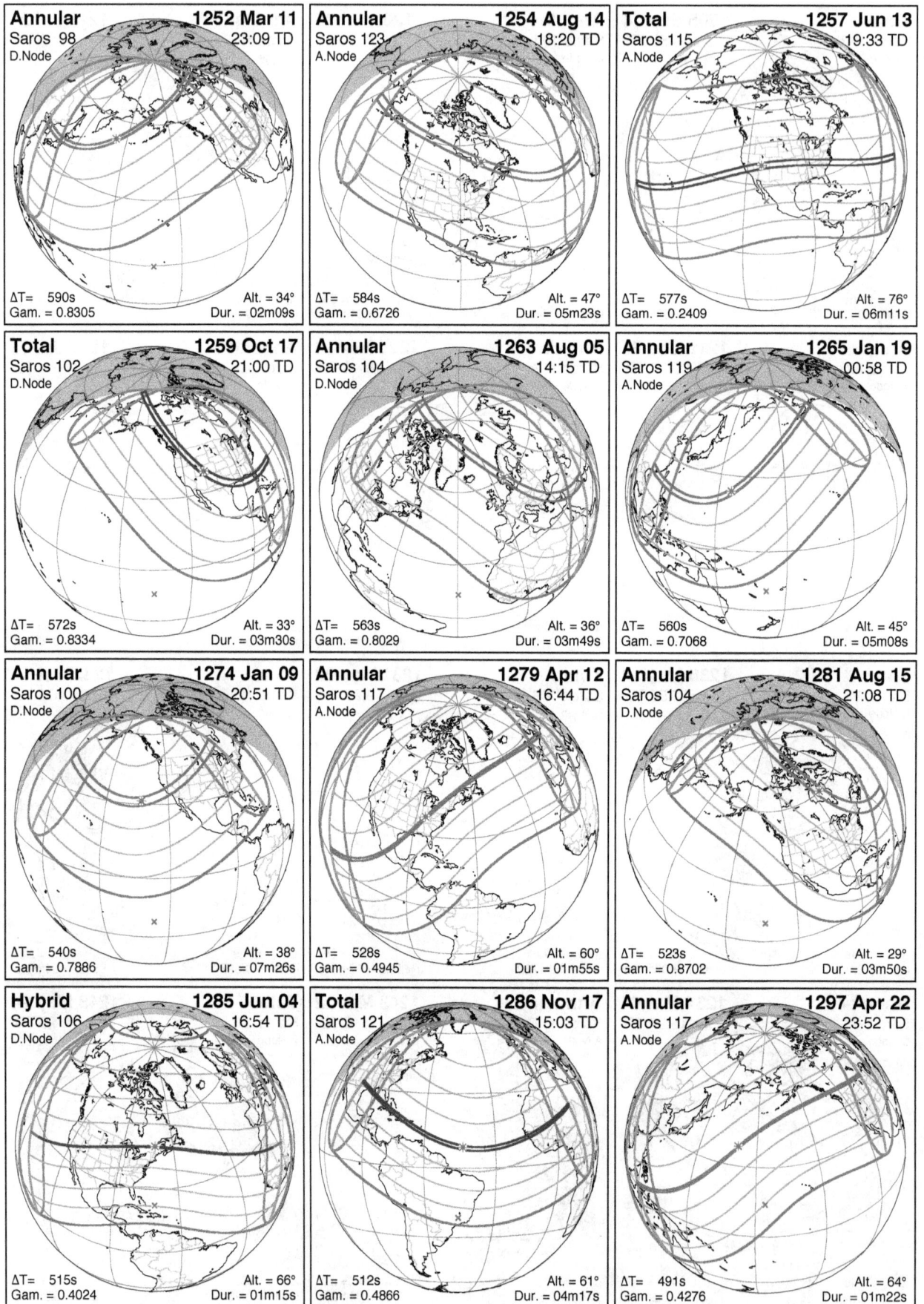

Annular 1252 Mar 11	
Saros 98 23:09 TD	
D.Node	
ΔT= 590s Alt. = 34°	
Gam. = 0.8305 Dur. = 02m09s	

Annular 1252 Mar 11
Saros 98
D.Node
23:09 TD
ΔT= 590s
Gam. = 0.8305
Alt. = 34°
Dur. = 02m09s

Annular 1254 Aug 14
Saros 123
A.Node
18:20 TD
ΔT= 584s
Gam. = 0.6726
Alt. = 47°
Dur. = 05m23s

Total 1257 Jun 13
Saros 115
A.Node
19:33 TD
ΔT= 577s
Gam. = 0.2409
Alt. = 76°
Dur. = 06m11s

Total 1259 Oct 17
Saros 102
D.Node
21:00 TD
ΔT= 572s
Gam. = 0.8334
Alt. = 33°
Dur. = 03m30s

Annular 1263 Aug 05
Saros 104
D.Node
14:15 TD
ΔT= 563s
Gam. = 0.8029
Alt. = 36°
Dur. = 03m49s

Annular 1265 Jan 19
Saros 119
A.Node
00:58 TD
ΔT= 560s
Gam. = 0.7068
Alt. = 45°
Dur. = 05m08s

Annular 1274 Jan 09
Saros 100
D.Node
20:51 TD
ΔT= 540s
Gam. = 0.7886
Alt. = 38°
Dur. = 07m26s

Annular 1279 Apr 12
Saros 117
A.Node
16:44 TD
ΔT= 528s
Gam. = 0.4945
Alt. = 60°
Dur. = 01m55s

Annular 1281 Aug 15
Saros 104
D.Node
21:08 TD
ΔT= 523s
Gam. = 0.8702
Alt. = 29°
Dur. = 03m50s

Hybrid 1285 Jun 04
Saros 106
D.Node
16:54 TD
ΔT= 515s
Gam. = 0.4024
Alt. = 66°
Dur. = 01m15s

Total 1286 Nov 17
Saros 121
A.Node
15:03 TD
ΔT= 512s
Gam. = 0.4866
Alt. = 61°
Dur. = 04m17s

Annular 1297 Apr 22
Saros 117
A.Node
23:52 TD
ΔT= 491s
Gam. = 0.4276
Alt. = 64°
Dur. = 01m22s

Plate 006

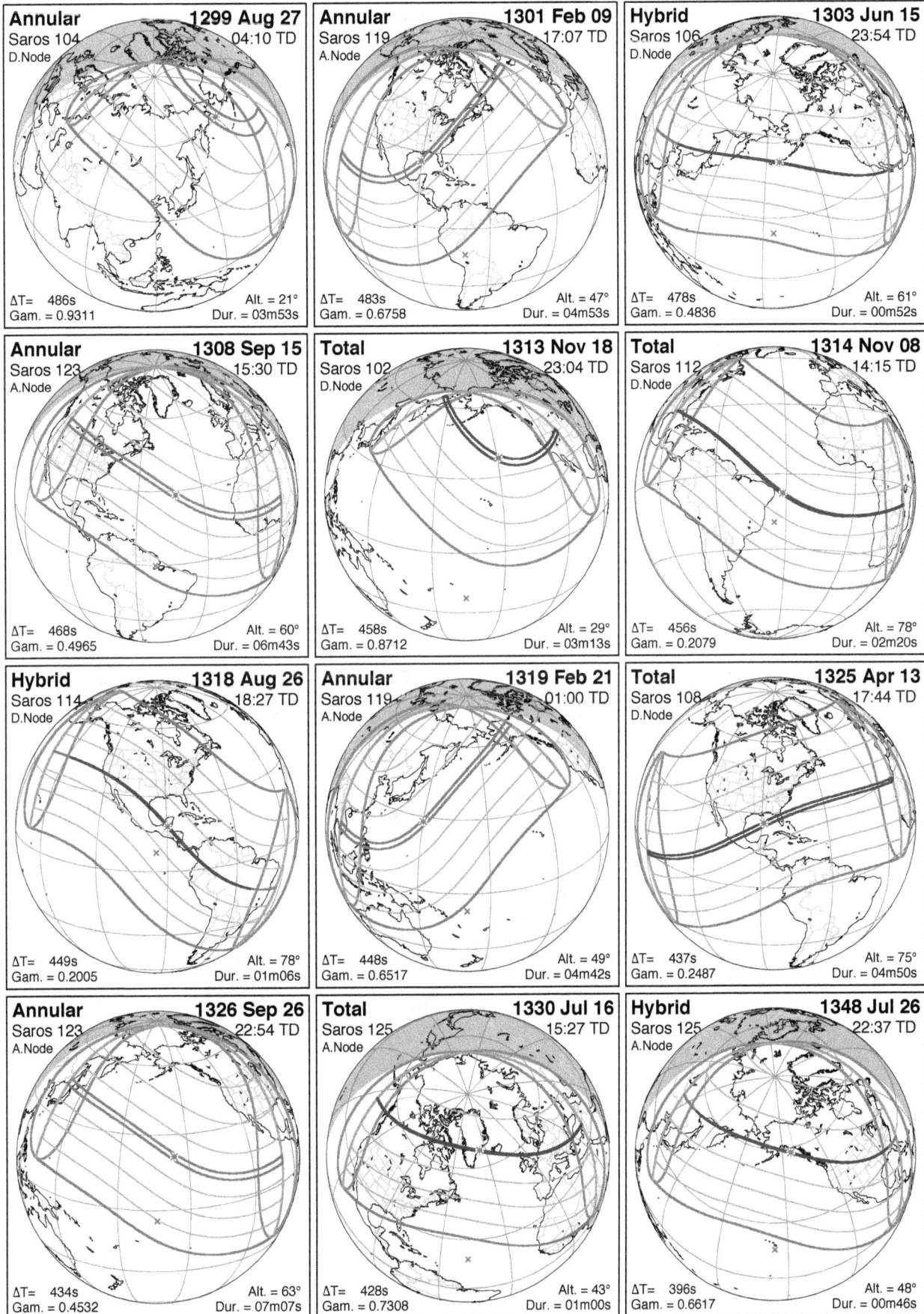

Annular	1299 Aug 27
Saros 104	04:10 TD
D.Node	
ΔT= 486s	Alt. = 21°
Gam. = 0.9311	Dur. = 03m53s

Annular	1301 Feb 09
Saros 119	17:07 TD
A.Node	
ΔT= 483s	Alt. = 47°
Gam. = 0.6758	Dur. = 04m53s

Hybrid	1303 Jun 15
Saros 106	23:54 TD
D.Node	
ΔT= 478s	Alt. = 61°
Gam. = 0.4836	Dur. = 00m52s

Annular	1308 Sep 15
Saros 123	15:30 TD
A.Node	
ΔT= 468s	Alt. = 60°
Gam. = 0.4965	Dur. = 06m43s

Total	1313 Nov 18
Saros 102	23:04 TD
D.Node	
ΔT= 458s	Alt. = 29°
Gam. = 0.8712	Dur. = 03m13s

Total	1314 Nov 08
Saros 112	14:15 TD
D.Node	
ΔT= 456s	Alt. = 78°
Gam. = 0.2079	Dur. = 02m20s

Hybrid	1318 Aug 26
Saros 114	18:27 TD
D.Node	
ΔT= 449s	Alt. = 78°
Gam. = 0.2005	Dur. = 01m06s

Annular	1319 Feb 21
Saros 119	01:00 TD
A.Node	
ΔT= 448s	Alt. = 49°
Gam. = 0.6517	Dur. = 04m42s

Total	1325 Apr 13
Saros 108	17:44 TD
D.Node	
ΔT= 437s	Alt. = 75°
Gam. = 0.2487	Dur. = 04m50s

Annular	1326 Sep 26
Saros 123	22:54 TD
A.Node	
ΔT= 434s	Alt. = 63°
Gam. = 0.4532	Dur. = 07m07s

Total	1330 Jul 16
Saros 125	15:27 TD
A.Node	
ΔT= 428s	Alt. = 43°
Gam. = 0.7308	Dur. = 01m00s

Hybrid	1348 Jul 26
Saros 125	22:37 TD
A.Node	
ΔT= 396s	Alt. = 48°
Gam. = 0.6617	Dur. = 00m46s

Plate 007

45

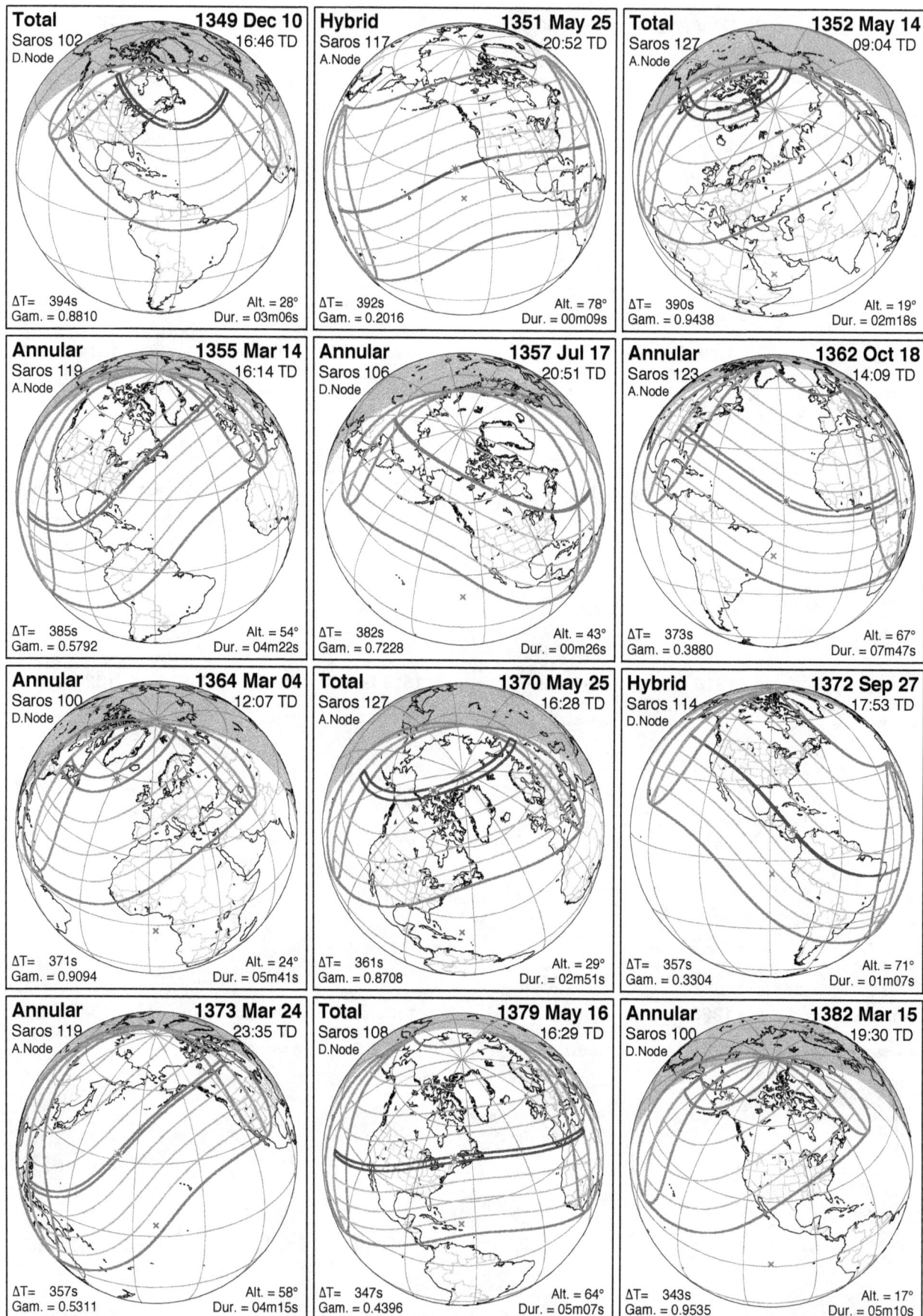

Total	1349 Dec 10
Saros 102	16:46 TD
D.Node	

ΔT= 394s   Alt. = 28°
Gam. = 0.8810   Dur. = 03m06s

Hybrid	1351 May 25
Saros 117	20:52 TD
A.Node	

ΔT= 392s   Alt. = 78°
Gam. = 0.2016   Dur. = 00m09s

Total	1352 May 14
Saros 127	09:04 TD
A.Node	

ΔT= 390s   Alt. = 19°
Gam. = 0.9438   Dur. = 02m18s

Annular	1355 Mar 14
Saros 119	16:14 TD
A.Node	

ΔT= 385s   Alt. = 54°
Gam. = 0.5792   Dur. = 04m22s

Annular	1357 Jul 17
Saros 106	20:51 TD
D.Node	

ΔT= 382s   Alt. = 43°
Gam. = 0.7228   Dur. = 00m26s

Annular	1362 Oct 18
Saros 123	14:09 TD
A.Node	

ΔT= 373s   Alt. = 67°
Gam. = 0.3880   Dur. = 07m47s

Annular	1364 Mar 04
Saros 100	12:07 TD
D.Node	

ΔT= 371s   Alt. = 24°
Gam. = 0.9094   Dur. = 05m41s

Total	1370 May 25
Saros 127	16:28 TD
A.Node	

ΔT= 361s   Alt. = 29°
Gam. = 0.8708   Dur. = 02m51s

Hybrid	1372 Sep 27
Saros 114	17:53 TD
D.Node	

ΔT= 357s   Alt. = 71°
Gam. = 0.3304   Dur. = 01m07s

Annular	1373 Mar 24
Saros 119	23:35 TD
A.Node	

ΔT= 357s   Alt. = 58°
Gam. = 0.5311   Dur. = 04m15s

Total	1379 May 16
Saros 108	16:29 TD
D.Node	

ΔT= 347s   Alt. = 64°
Gam. = 0.4396   Dur. = 05m07s

Annular	1382 Mar 15
Saros 100	19:30 TD
D.Node	

ΔT= 343s   Alt. = 17°
Gam. = 0.9535   Dur. = 05m10s

Plate 008

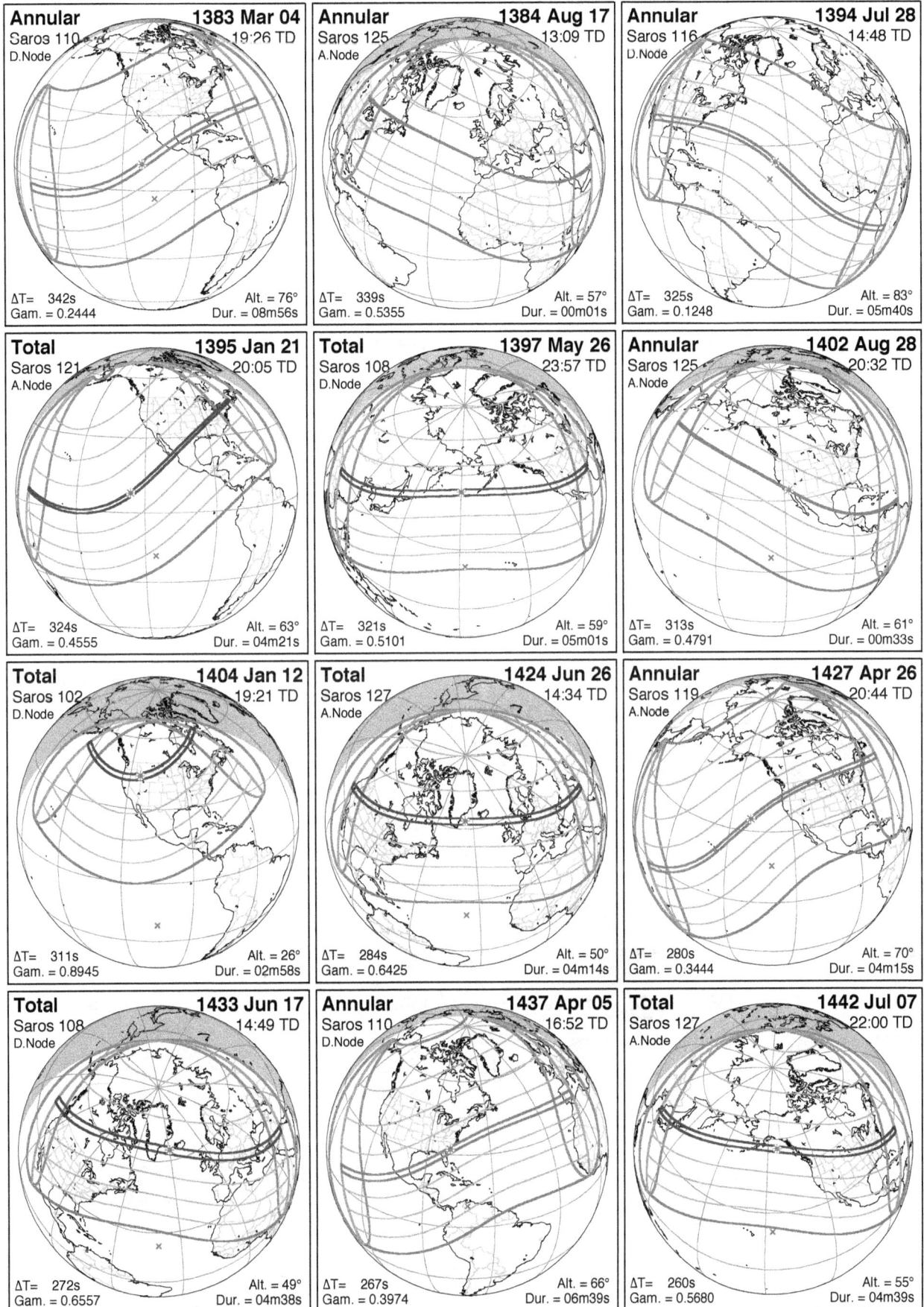

Annular	1383 Mar 04
Saros 110	19:26 TD
D.Node	
ΔT= 342s	Alt. = 76°
Gam. = 0.2444	Dur. = 08m56s

Annular	1384 Aug 17
Saros 125	13:09 TD
A.Node	
ΔT= 339s	Alt. = 57°
Gam. = 0.5355	Dur. = 00m01s

Annular	1394 Jul 28
Saros 116	14:48 TD
D.Node	
ΔT= 325s	Alt. = 83°
Gam. = 0.1248	Dur. = 05m40s

Total	1395 Jan 21
Saros 121	20:05 TD
A.Node	
ΔT= 324s	Alt. = 63°
Gam. = 0.4555	Dur. = 04m21s

Total	1397 May 26
Saros 108	23:57 TD
D.Node	
ΔT= 321s	Alt. = 59°
Gam. = 0.5101	Dur. = 05m01s

Annular	1402 Aug 28
Saros 125	20:32 TD
A.Node	
ΔT= 313s	Alt. = 61°
Gam. = 0.4791	Dur. = 00m33s

Total	1404 Jan 12
Saros 102	19:21 TD
D.Node	
ΔT= 311s	Alt. = 26°
Gam. = 0.8945	Dur. = 02m58s

Total	1424 Jun 26
Saros 127	14:34 TD
A.Node	
ΔT= 284s	Alt. = 50°
Gam. = 0.6425	Dur. = 04m14s

Annular	1427 Apr 26
Saros 119	20:44 TD
A.Node	
ΔT= 280s	Alt. = 70°
Gam. = 0.3444	Dur. = 04m15s

Total	1433 Jun 17
Saros 108	14:49 TD
D.Node	
ΔT= 272s	Alt. = 49°
Gam. = 0.6557	Dur. = 04m38s

Annular	1437 Apr 05
Saros 110	16:52 TD
D.Node	
ΔT= 267s	Alt. = 66°
Gam. = 0.3974	Dur. = 06m39s

Total	1442 Jul 07
Saros 127	22:00 TD
A.Node	
ΔT= 260s	Alt. = 55°
Gam. = 0.5680	Dur. = 04m39s

Plate 009

47

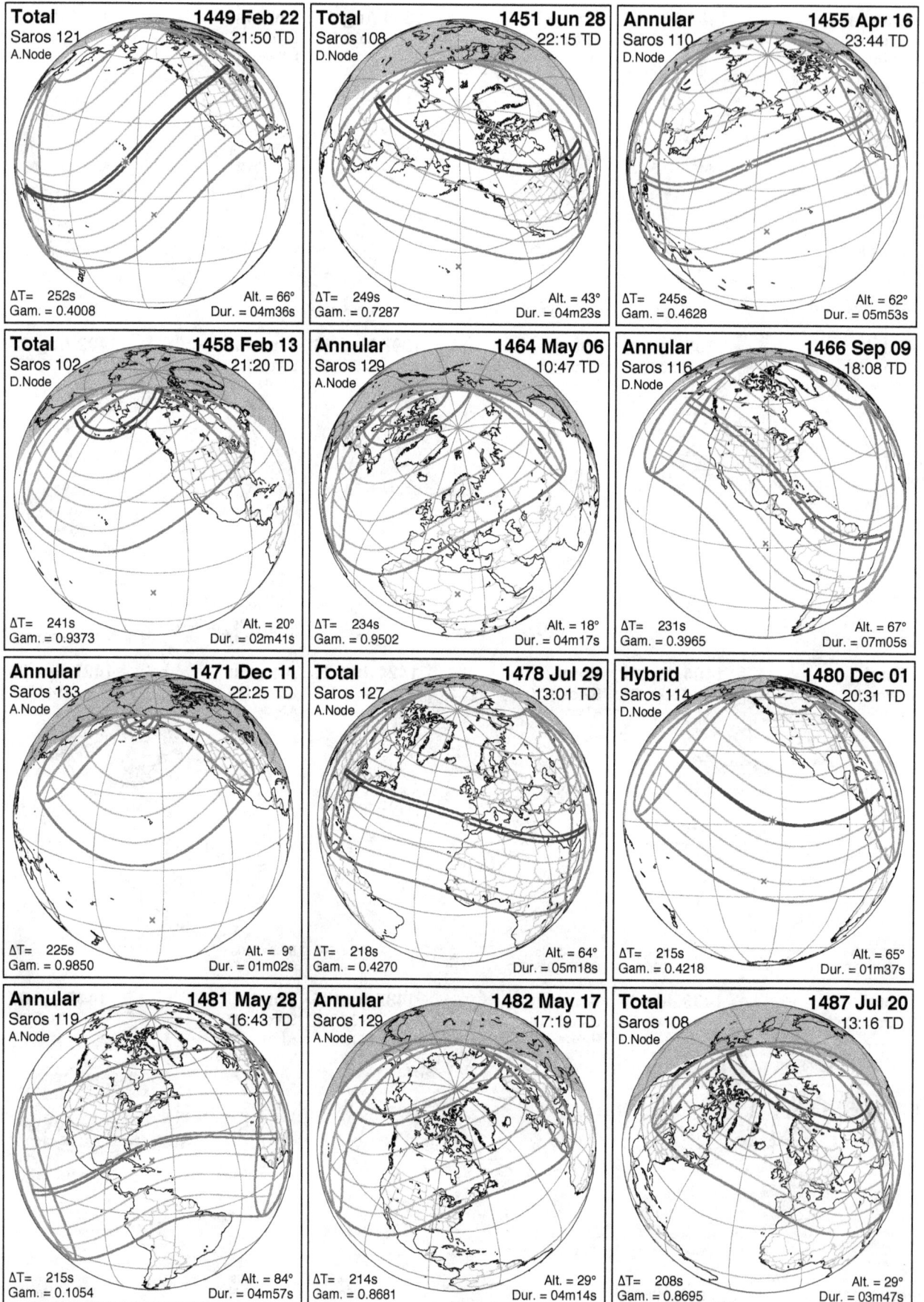

Total	1449 Feb 22
Saros 121	21:50 TD
A.Node	
ΔT= 252s	Alt. = 66°
Gam. = 0.4008	Dur. = 04m36s

Total	1451 Jun 28
Saros 108	22:15 TD
D.Node	
ΔT= 249s	Alt. = 43°
Gam. = 0.7287	Dur. = 04m23s

Annular	1455 Apr 16
Saros 110	23:44 TD
D.Node	
ΔT= 245s	Alt. = 62°
Gam. = 0.4628	Dur. = 05m53s

Total	1458 Feb 13
Saros 102	21:20 TD
D.Node	
ΔT= 241s	Alt. = 20°
Gam. = 0.9373	Dur. = 02m41s

Annular	1464 May 06
Saros 129	10:47 TD
A.Node	
ΔT= 234s	Alt. = 18°
Gam. = 0.9502	Dur. = 04m17s

Annular	1466 Sep 09
Saros 116	18:08 TD
D.Node	
ΔT= 231s	Alt. = 67°
Gam. = 0.3965	Dur. = 07m05s

Annular	1471 Dec 11
Saros 133	22:25 TD
A.Node	
ΔT= 225s	Alt. = 9°
Gam. = 0.9850	Dur. = 01m02s

Total	1478 Jul 29
Saros 127	13:01 TD
A.Node	
ΔT= 218s	Alt. = 64°
Gam. = 0.4270	Dur. = 05m18s

Hybrid	1480 Dec 01
Saros 114	20:31 TD
D.Node	
ΔT= 215s	Alt. = 65°
Gam. = 0.4218	Dur. = 01m37s

Annular	1481 May 28
Saros 119	16:43 TD
A.Node	
ΔT= 215s	Alt. = 84°
Gam. = 0.1054	Dur. = 04m57s

Annular	1482 May 17
Saros 129	17:19 TD
A.Node	
ΔT= 214s	Alt. = 29°
Gam. = 0.8681	Dur. = 04m14s

Total	1487 Jul 20
Saros 108	13:16 TD
D.Node	
ΔT= 208s	Alt. = 29°
Gam. = 0.8695	Dur. = 03m47s

Plate 010

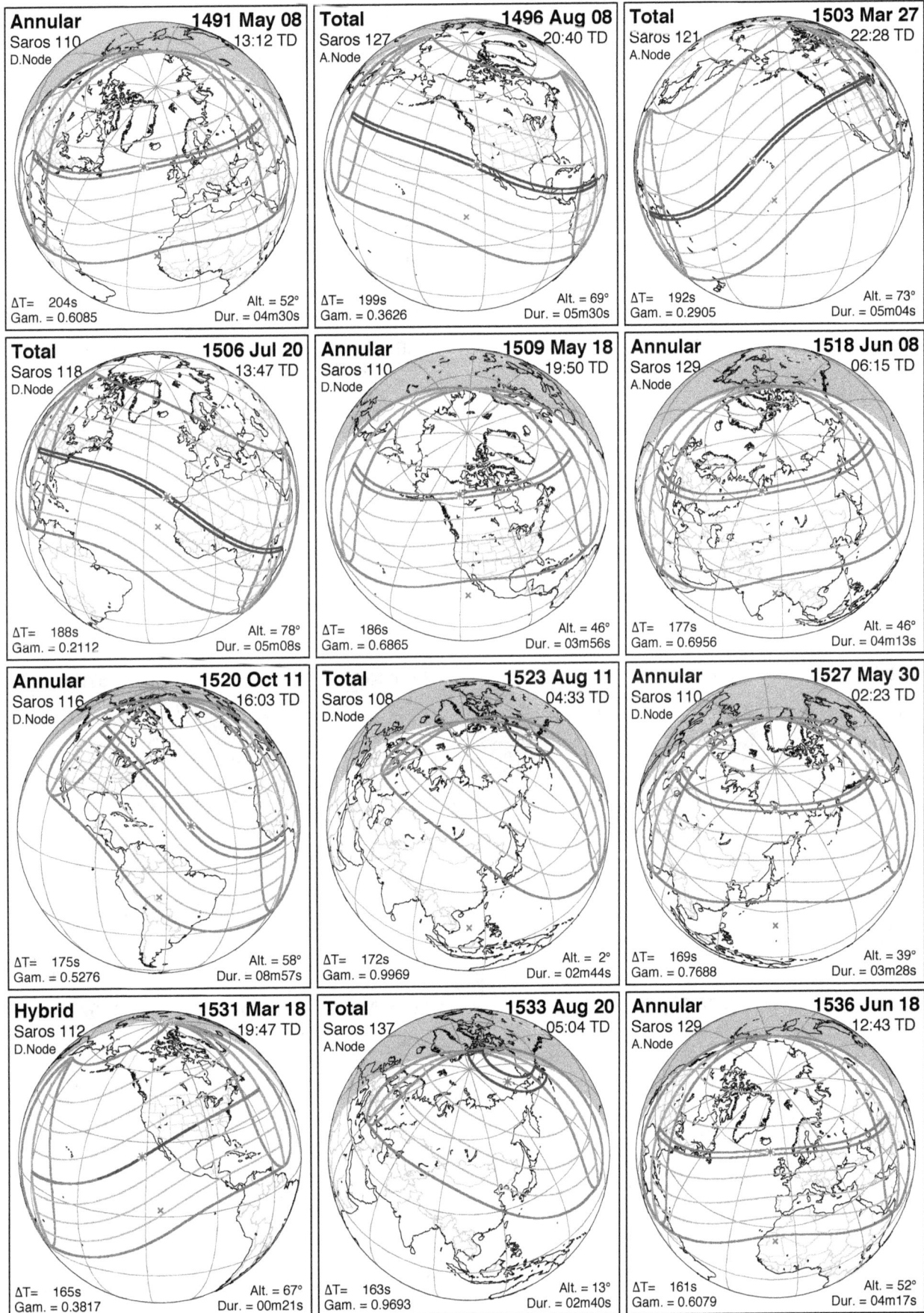

Annular **1491 May 08**	Saros 110, D.Node, 13:12 TD

Annular   **1491 May 08**
Saros 110
D.Node   13:12 TD
ΔT= 204s   Alt. = 52°
Gam. = 0.6085   Dur. = 04m30s

Total   **1496 Aug 08**
Saros 127
A.Node   20:40 TD
ΔT= 199s   Alt. = 69°
Gam. = 0.3626   Dur. = 05m30s

Total   **1503 Mar 27**
Saros 121
A.Node   22:28 TD
ΔT= 192s   Alt. = 73°
Gam. = 0.2905   Dur. = 05m04s

Total   **1506 Jul 20**
Saros 118
D.Node   13:47 TD
ΔT= 188s   Alt. = 78°
Gam. = 0.2112   Dur. = 05m08s

Annular   **1509 May 18**
Saros 110
D.Node   19:50 TD
ΔT= 186s   Alt. = 46°
Gam. = 0.6865   Dur. = 03m56s

Annular   **1518 Jun 08**
Saros 129
A.Node   06:15 TD
ΔT= 177s   Alt. = 46°
Gam. = 0.6956   Dur. = 04m13s

Annular   **1520 Oct 11**
Saros 116
D.Node   16:03 TD
ΔT= 175s   Alt. = 58°
Gam. = 0.5276   Dur. = 08m57s

Total   **1523 Aug 11**
Saros 108
D.Node   04:33 TD
ΔT= 172s   Alt. = 2°
Gam. = 0.9969   Dur. = 02m44s

Annular   **1527 May 30**
Saros 110
D.Node   02:23 TD
ΔT= 169s   Alt. = 39°
Gam. = 0.7688   Dur. = 03m28s

Hybrid   **1531 Mar 18**
Saros 112
D.Node   19:47 TD
ΔT= 165s   Alt. = 67°
Gam. = 0.3817   Dur. = 00m21s

Total   **1533 Aug 20**
Saros 137
A.Node   05:04 TD
ΔT= 163s   Alt. = 13°
Gam. = 0.9693   Dur. = 02m40s

Annular   **1536 Jun 18**
Saros 129
A.Node   12:43 TD
ΔT= 161s   Alt. = 52°
Gam. = 0.6079   Dur. = 04m17s

Plate 011

49

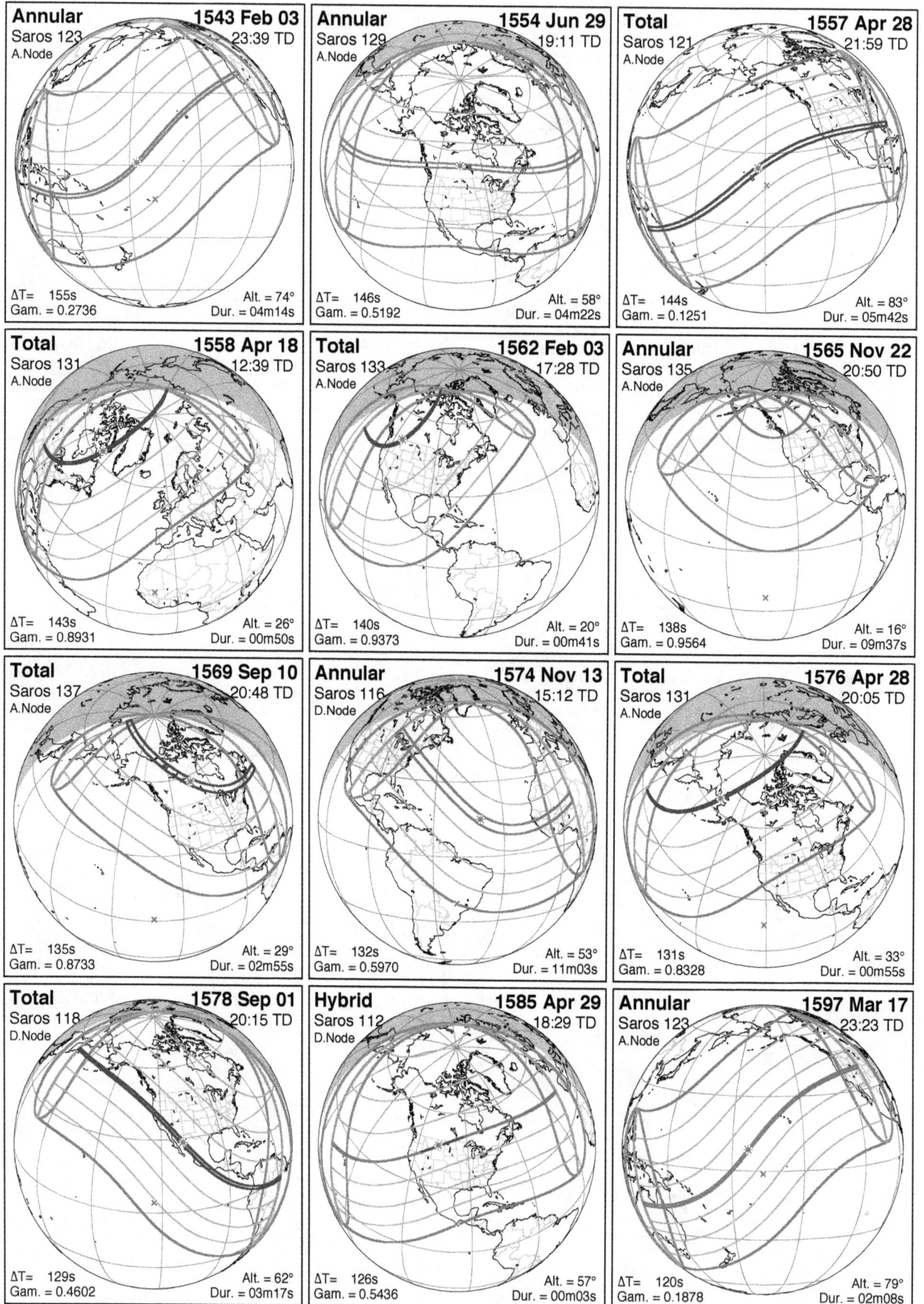

Annular **1543 Feb 03**	**Annular** **1554 Jun 29**
Saros 123 23:39 TD	Saros 129 19:11 TD
A.Node	A.Node

Annular   **1543 Feb 03**
Saros 123   23:39 TD
A.Node
ΔT= 155s   Alt. = 74°
Gam. = 0.2736   Dur. = 04m14s

Annular   **1554 Jun 29**
Saros 129   19:11 TD
A.Node
ΔT= 146s   Alt. = 58°
Gam. = 0.5192   Dur. = 04m22s

Total   **1557 Apr 28**
Saros 121   21:59 TD
A.Node
ΔT= 144s   Alt. = 83°
Gam. = 0.1251   Dur. = 05m42s

Total   **1558 Apr 18**
Saros 131   12:39 TD
A.Node
ΔT= 143s   Alt. = 26°
Gam. = 0.8931   Dur. = 00m50s

Total   **1562 Feb 03**
Saros 133   17:28 TD
A.Node
ΔT= 140s   Alt. = 20°
Gam. = 0.9373   Dur. = 00m41s

Annular   **1565 Nov 22**
Saros 135   20:50 TD
A.Node
ΔT= 138s   Alt. = 16°
Gam. = 0.9564   Dur. = 09m37s

Total   **1569 Sep 10**
Saros 137   20:48 TD
A.Node
ΔT= 135s   Alt. = 29°
Gam. = 0.8733   Dur. = 02m55s

Annular   **1574 Nov 13**
Saros 116   15:12 TD
D.Node
ΔT= 132s   Alt. = 53°
Gam. = 0.5970   Dur. = 11m03s

Total   **1576 Apr 28**
Saros 131   20:05 TD
A.Node
ΔT= 131s   Alt. = 33°
Gam. = 0.8328   Dur. = 00m55s

Total   **1578 Sep 01**
Saros 118   20:15 TD
D.Node
ΔT= 129s   Alt. = 62°
Gam. = 0.4602   Dur. = 03m17s

Hybrid   **1585 Apr 29**
Saros 112   18:29 TD
D.Node
ΔT= 126s   Alt. = 57°
Gam. = 0.5436   Dur. = 00m03s

Annular   **1597 Mar 17**
Saros 123   23:23 TD
A.Node
ΔT= 120s   Alt. = 79°
Gam. = 0.1878   Dur. = 02m08s

Plate 012

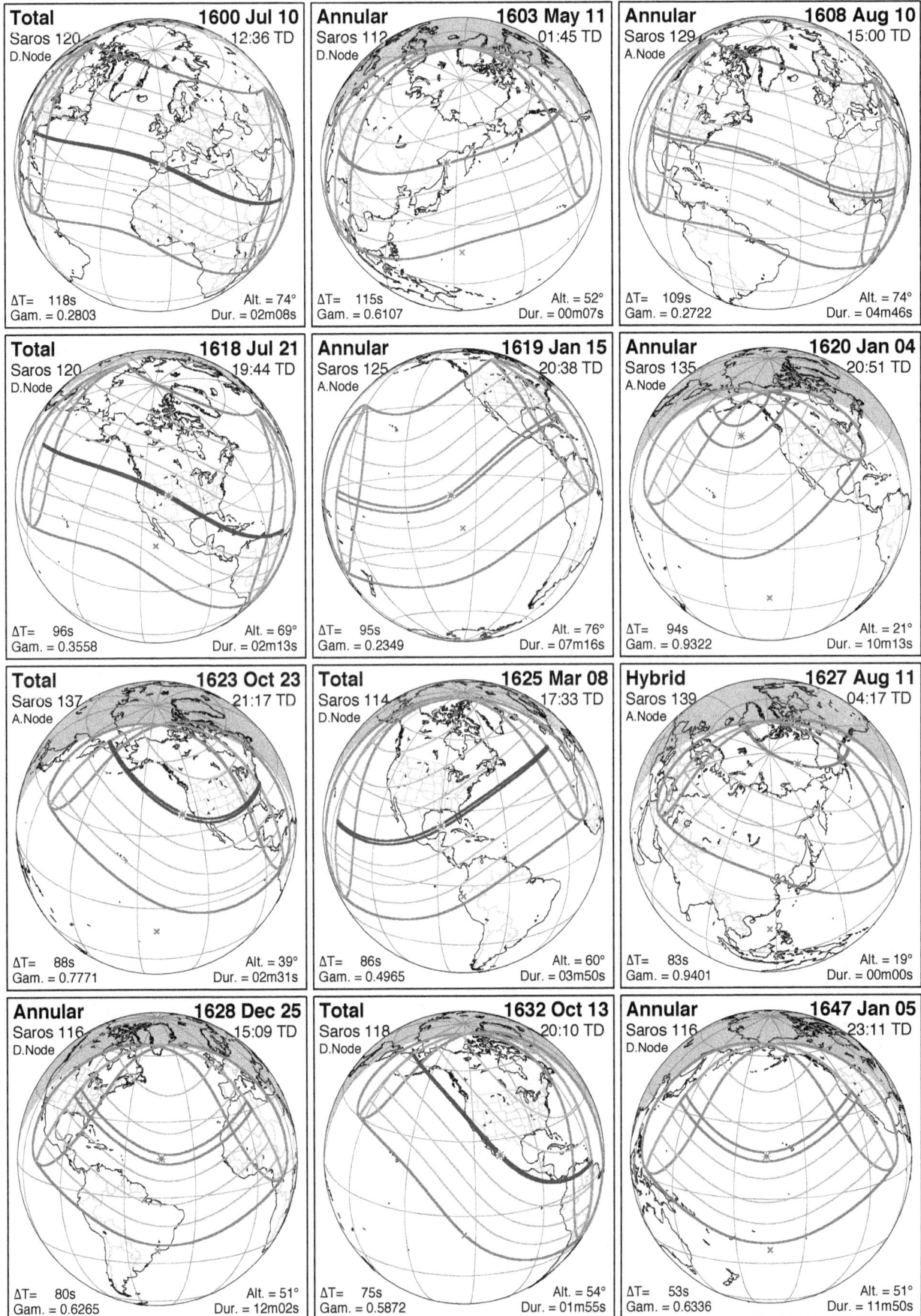

Total 1600 Jul 10	Annular 1603 May 11	Annular 1608 Aug 10
Saros 120, D.Node, 12:36 TD, ΔT= 118s, Gam. = 0.2803, Alt. = 74°, Dur. = 02m08s	Saros 112, D.Node, 01:45 TD, ΔT= 115s, Gam. = 0.6107, Alt. = 52°, Dur. = 00m07s	Saros 129, A.Node, 15:00 TD, ΔT= 109s, Gam. = 0.2722, Alt. = 74°, Dur. = 04m46s
Total 1618 Jul 21	Annular 1619 Jan 15	Annular 1620 Jan 04
Saros 120, D.Node, 19:44 TD, ΔT= 96s, Gam. = 0.3558, Alt. = 69°, Dur. = 02m13s	Saros 125, A.Node, 20:38 TD, ΔT= 95s, Gam. = 0.2349, Alt. = 76°, Dur. = 07m16s	Saros 135, A.Node, 20:51 TD, ΔT= 94s, Gam. = 0.9322, Alt. = 21°, Dur. = 10m13s
Total 1623 Oct 23	Total 1625 Mar 08	Hybrid 1627 Aug 11
Saros 137, A.Node, 21:17 TD, ΔT= 88s, Gam. = 0.7771, Alt. = 39°, Dur. = 02m31s	Saros 114, D.Node, 17:33 TD, ΔT= 86s, Gam. = 0.4965, Alt. = 60°, Dur. = 03m50s	Saros 139, A.Node, 04:17 TD, ΔT= 83s, Gam. = 0.9401, Alt. = 19°, Dur. = 00m00s
Annular 1628 Dec 25	Total 1632 Oct 13	Annular 1647 Jan 05
Saros 116, D.Node, 15:09 TD, ΔT= 80s, Gam. = 0.6265, Alt. = 51°, Dur. = 12m02s	Saros 118, D.Node, 20:10 TD, ΔT= 75s, Gam. = 0.5872, Alt. = 54°, Dur. = 01m55s	Saros 116, D.Node, 23:11 TD, ΔT= 53s, Gam. = 0.6336, Alt. = 51°, Dur. = 11m50s

Plate 013

51

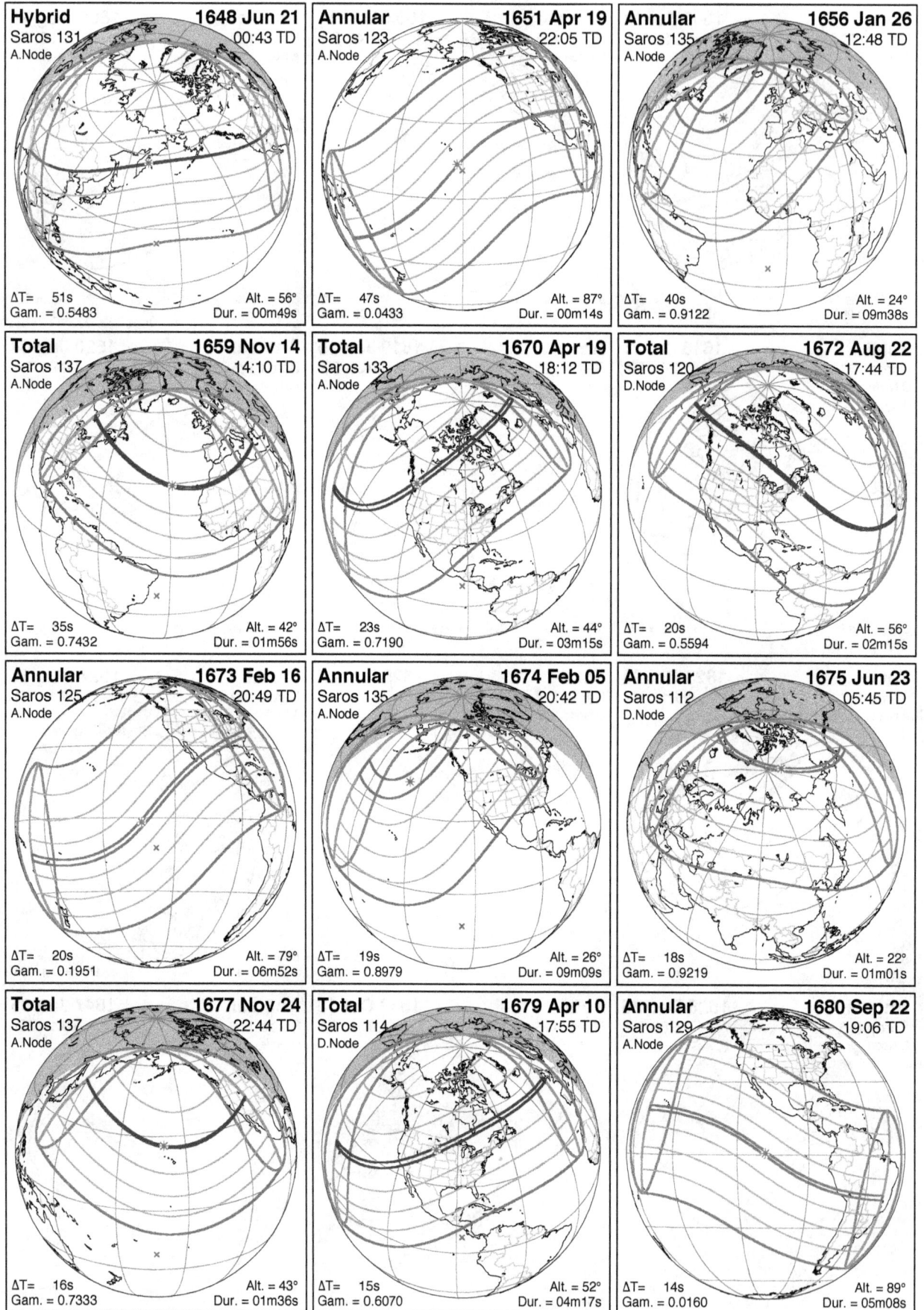

Hybrid **1648 Jun 21**	
Saros 131 00:43 TD	
A.Node	
ΔT= 51s	Alt. = 56°
Gam. = 0.5483	Dur. = 00m49s

Annular **1651 Apr 19**	
Saros 123 22:05 TD	
A.Node	
ΔT= 47s	Alt. = 87°
Gam. = 0.0433	Dur. = 00m14s

Annular **1656 Jan 26**	
Saros 135 12:48 TD	
A.Node	
ΔT= 40s	Alt. = 24°
Gam. = 0.9122	Dur. = 09m38s

Total **1659 Nov 14**	
Saros 137 14:10 TD	
A.Node	
ΔT= 35s	Alt. = 42°
Gam. = 0.7432	Dur. = 01m56s

Total **1670 Apr 19**	
Saros 133 18:12 TD	
A.Node	
ΔT= 23s	Alt. = 44°
Gam. = 0.7190	Dur. = 03m15s

Total **1672 Aug 22**	
Saros 120 17:44 TD	
D.Node	
ΔT= 20s	Alt. = 56°
Gam. = 0.5594	Dur. = 02m15s

Annular **1673 Feb 16**	
Saros 125 20:49 TD	
A.Node	
ΔT= 20s	Alt. = 79°
Gam. = 0.1951	Dur. = 06m52s

Annular **1674 Feb 05**	
Saros 135 20:42 TD	
A.Node	
ΔT= 19s	Alt. = 26°
Gam. = 0.8979	Dur. = 09m09s

Annular **1675 Jun 23**	
Saros 112 05:45 TD	
D.Node	
ΔT= 18s	Alt. = 22°
Gam. = 0.9219	Dur. = 01m01s

Total **1677 Nov 24**	
Saros 137 22:44 TD	
A.Node	
ΔT= 16s	Alt. = 43°
Gam. = 0.7333	Dur. = 01m36s

Total **1679 Apr 10**	
Saros 114 17:55 TD	
D.Node	
ΔT= 15s	Alt. = 52°
Gam. = 0.6070	Dur. = 04m17s

Annular **1680 Sep 22**	
Saros 129 19:06 TD	
A.Node	
ΔT= 14s	Alt. = 89°
Gam. = 0.0160	Dur. = 05m08s

Plate 014

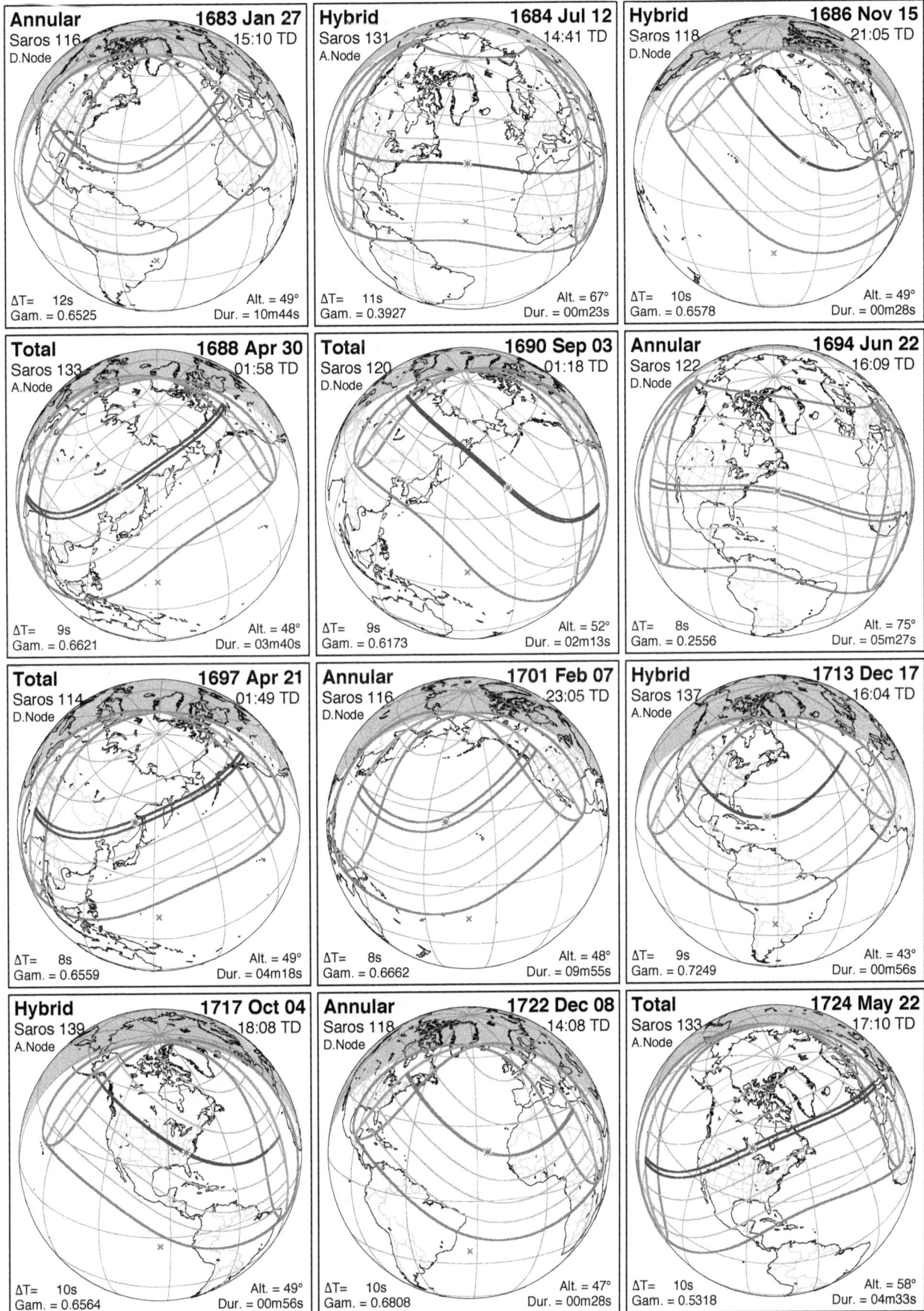

Annular	1683 Jan 27
Saros 116	15:10 TD
D.Node	
ΔT= 12s	Alt. = 49°
Gam. = 0.6525	Dur. = 10m44s

Hybrid	1684 Jul 12
Saros 131	14:41 TD
A.Node	
ΔT= 11s	Alt. = 67°
Gam. = 0.3927	Dur. = 00m23s

Hybrid	1686 Nov 15
Saros 118	21:05 TD
D.Node	
ΔT= 10s	Alt. = 49°
Gam. = 0.6578	Dur. = 00m28s

Total	1688 Apr 30
Saros 133	01:58 TD
A.Node	
ΔT= 9s	Alt. = 48°
Gam. = 0.6621	Dur. = 03m40s

Total	1690 Sep 03
Saros 120	01:18 TD
D.Node	
ΔT= 9s	Alt. = 52°
Gam. = 0.6173	Dur. = 02m13s

Annular	1694 Jun 22
Saros 122	16:09 TD
D.Node	
ΔT= 8s	Alt. = 75°
Gam. = 0.2556	Dur. = 05m27s

Total	1697 Apr 21
Saros 114	01:49 TD
D.Node	
ΔT= 8s	Alt. = 49°
Gam. = 0.6559	Dur. = 04m18s

Annular	1701 Feb 07
Saros 116	23:05 TD
D.Node	
ΔT= 8s	Alt. = 48°
Gam. = 0.6662	Dur. = 09m55s

Hybrid	1713 Dec 17
Saros 137	16:04 TD
A.Node	
ΔT= 9s	Alt. = 43°
Gam. = 0.7249	Dur. = 00m56s

Hybrid	1717 Oct 04
Saros 139	18:08 TD
A.Node	
ΔT= 10s	Alt. = 49°
Gam. = 0.6564	Dur. = 00m56s

Annular	1722 Dec 08
Saros 118	14:08 TD
D.Node	
ΔT= 10s	Alt. = 47°
Gam. = 0.6808	Dur. = 00m28s

Total	1724 May 22
Saros 133	17:10 TD
A.Node	
ΔT= 10s	Alt. = 58°
Gam. = 0.5318	Dur. = 04m33s

Plate 015

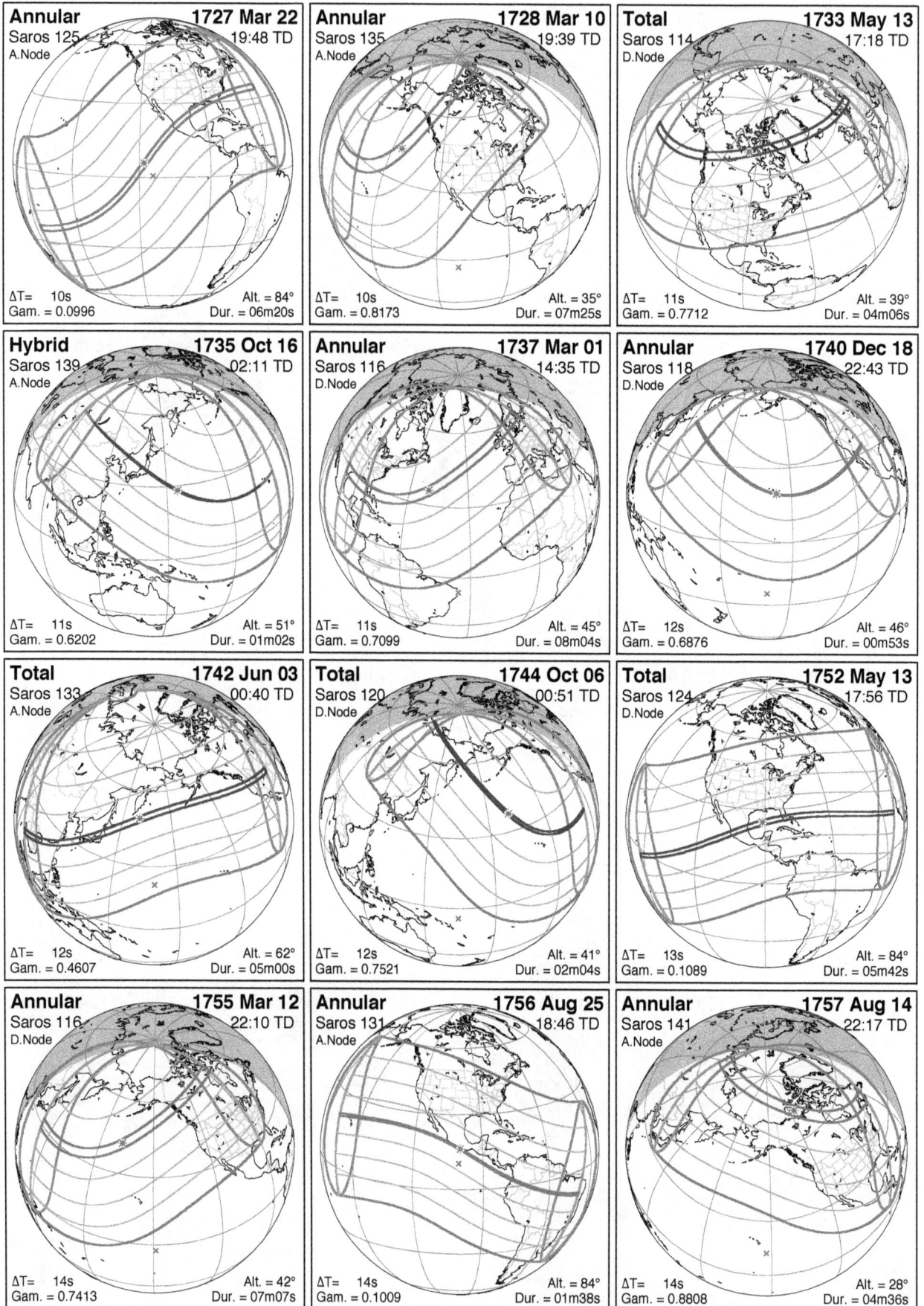

Annular **1727 Mar 22**	
Saros 125 19:48 TD	
A.Node	
ΔT= 10s	Alt. = 84°
Gam. = 0.0996	Dur. = 06m20s

Annular **1728 Mar 10**	
Saros 135 19:39 TD	
A.Node	
ΔT= 10s	Alt. = 35°
Gam. = 0.8173	Dur. = 07m25s

Total **1733 May 13**	
Saros 114 17:18 TD	
D.Node	
ΔT= 11s	Alt. = 39°
Gam. = 0.7712	Dur. = 04m06s

Hybrid **1735 Oct 16**	
Saros 139 02:11 TD	
A.Node	
ΔT= 11s	Alt. = 51°
Gam. = 0.6202	Dur. = 01m02s

Annular **1737 Mar 01**	
Saros 116 14:35 TD	
D.Node	
ΔT= 11s	Alt. = 45°
Gam. = 0.7099	Dur. = 08m04s

Annular **1740 Dec 18**	
Saros 118 22:43 TD	
D.Node	
ΔT= 12s	Alt. = 46°
Gam. = 0.6876	Dur. = 00m53s

Total **1742 Jun 03**	
Saros 133 00:40 TD	
A.Node	
ΔT= 12s	Alt. = 62°
Gam. = 0.4607	Dur. = 05m00s

Total **1744 Oct 06**	
Saros 120 00:51 TD	
D.Node	
ΔT= 12s	Alt. = 41°
Gam. = 0.7521	Dur. = 02m04s

Total **1752 May 13**	
Saros 124 17:56 TD	
D.Node	
ΔT= 13s	Alt. = 84°
Gam. = 0.1089	Dur. = 05m42s

Annular **1755 Mar 12**	
Saros 116 22:10 TD	
D.Node	
ΔT= 14s	Alt. = 42°
Gam. = 0.7413	Dur. = 07m07s

Annular **1756 Aug 25**	
Saros 131 18:46 TD	
A.Node	
ΔT= 14s	Alt. = 84°
Gam. = 0.1009	Dur. = 01m38s

Annular **1757 Aug 14**	
Saros 141 22:17 TD	
A.Node	
ΔT= 14s	Alt. = 28°
Gam. = 0.8808	Dur. = 04m36s

Plate 016

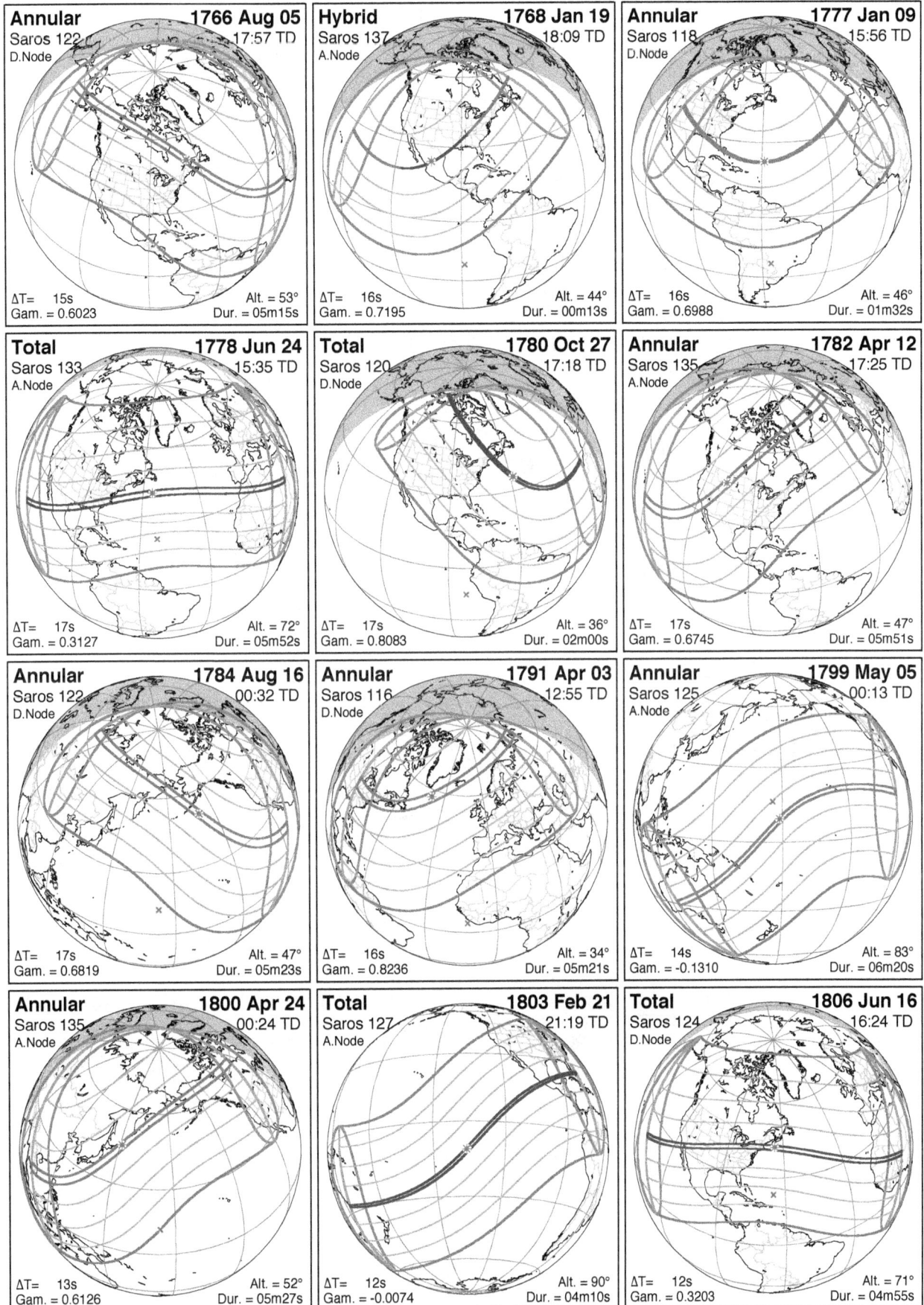

Annular	1766 Aug 05
Saros 122	17:57 TD
D.Node	
ΔT= 15s	Alt. = 53°
Gam. = 0.6023	Dur. = 05m15s

Hybrid	1768 Jan 19
Saros 137	18:09 TD
A.Node	
ΔT= 16s	Alt. = 44°
Gam. = 0.7195	Dur. = 00m13s

Annular	1777 Jan 09
Saros 118	15:56 TD
D.Node	
ΔT= 16s	Alt. = 46°
Gam. = 0.6988	Dur. = 01m32s

Total	1778 Jun 24
Saros 133	15:35 TD
A.Node	
ΔT= 17s	Alt. = 72°
Gam. = 0.3127	Dur. = 05m52s

Total	1780 Oct 27
Saros 120	17:18 TD
D.Node	
ΔT= 17s	Alt. = 36°
Gam. = 0.8083	Dur. = 02m00s

Annular	1782 Apr 12
Saros 135	17:25 TD
A.Node	
ΔT= 17s	Alt. = 47°
Gam. = 0.6745	Dur. = 05m51s

Annular	1784 Aug 16
Saros 122	00:32 TD
D.Node	
ΔT= 17s	Alt. = 47°
Gam. = 0.6819	Dur. = 05m23s

Annular	1791 Apr 03
Saros 116	12:55 TD
D.Node	
ΔT= 16s	Alt. = 34°
Gam. = 0.8236	Dur. = 05m21s

Annular	1799 May 05
Saros 125	00:13 TD
A.Node	
ΔT= 14s	Alt. = 83°
Gam. = -0.1310	Dur. = 06m20s

Annular	1800 Apr 24
Saros 135	00:24 TD
A.Node	
ΔT= 13s	Alt. = 52°
Gam. = 0.6126	Dur. = 05m27s

Total	1803 Feb 21
Saros 127	21:19 TD
A.Node	
ΔT= 12s	Alt. = 90°
Gam. = -0.0074	Dur. = 04m10s

Total	1806 Jun 16
Saros 124	16:24 TD
D.Node	
ΔT= 12s	Alt. = 71°
Gam. = 0.3203	Dur. = 04m55s

Plate 017

55

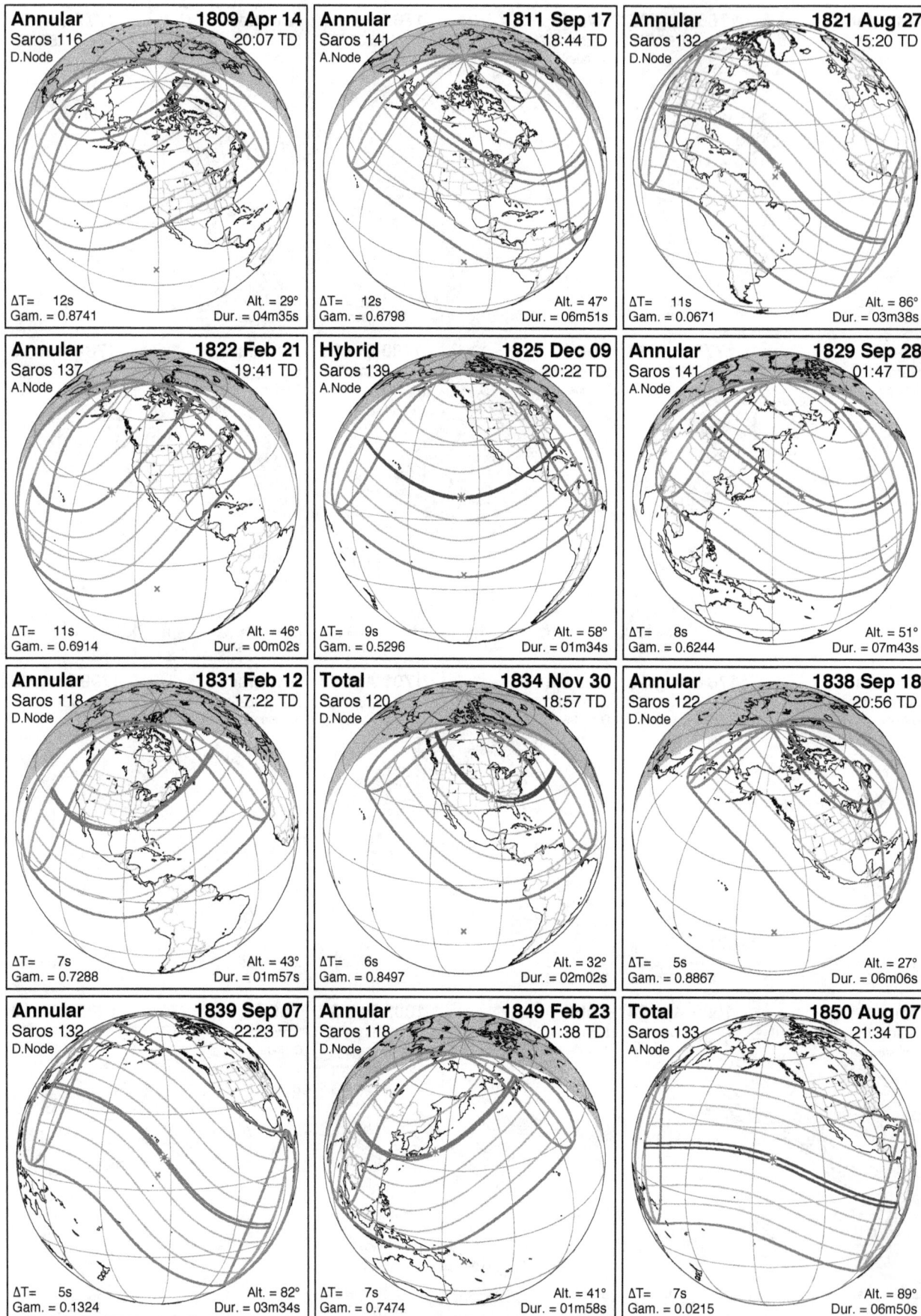

Annular	1809 Apr 14
Saros 116	20:07 TD
D.Node	
ΔT= 12s	Alt. = 29°
Gam. = 0.8741	Dur. = 04m35s

Annular	1811 Sep 17
Saros 141	18:44 TD
A.Node	
ΔT= 12s	Alt. = 47°
Gam. = 0.6798	Dur. = 06m51s

Annular	1821 Aug 27
Saros 132	15:20 TD
D.Node	
ΔT= 11s	Alt. = 86°
Gam. = 0.0671	Dur. = 03m38s

Annular	1822 Feb 21
Saros 137	19:41 TD
A.Node	
ΔT= 11s	Alt. = 46°
Gam. = 0.6914	Dur. = 00m02s

Hybrid	1825 Dec 09
Saros 139	20:22 TD
A.Node	
ΔT= 9s	Alt. = 58°
Gam. = 0.5296	Dur. = 01m34s

Annular	1829 Sep 28
Saros 141	01:47 TD
A.Node	
ΔT= 8s	Alt. = 51°
Gam. = 0.6244	Dur. = 07m43s

Annular	1831 Feb 12
Saros 118	17:22 TD
D.Node	
ΔT= 7s	Alt. = 43°
Gam. = 0.7288	Dur. = 01m57s

Total	1834 Nov 30
Saros 120	18:57 TD
D.Node	
ΔT= 6s	Alt. = 32°
Gam. = 0.8497	Dur. = 02m02s

Annular	1838 Sep 18
Saros 122	20:56 TD
D.Node	
ΔT= 5s	Alt. = 27°
Gam. = 0.8867	Dur. = 06m06s

Annular	1839 Sep 07
Saros 132	22:23 TD
D.Node	
ΔT= 5s	Alt. = 82°
Gam. = 0.1324	Dur. = 03m34s

Annular	1849 Feb 23
Saros 118	01:38 TD
D.Node	
ΔT= 7s	Alt. = 41°
Gam. = 0.7474	Dur. = 01m58s

Total	1850 Aug 07
Saros 133	21:34 TD
A.Node	
ΔT= 7s	Alt. = 89°
Gam. = 0.0215	Dur. = 06m50s

Plate 018

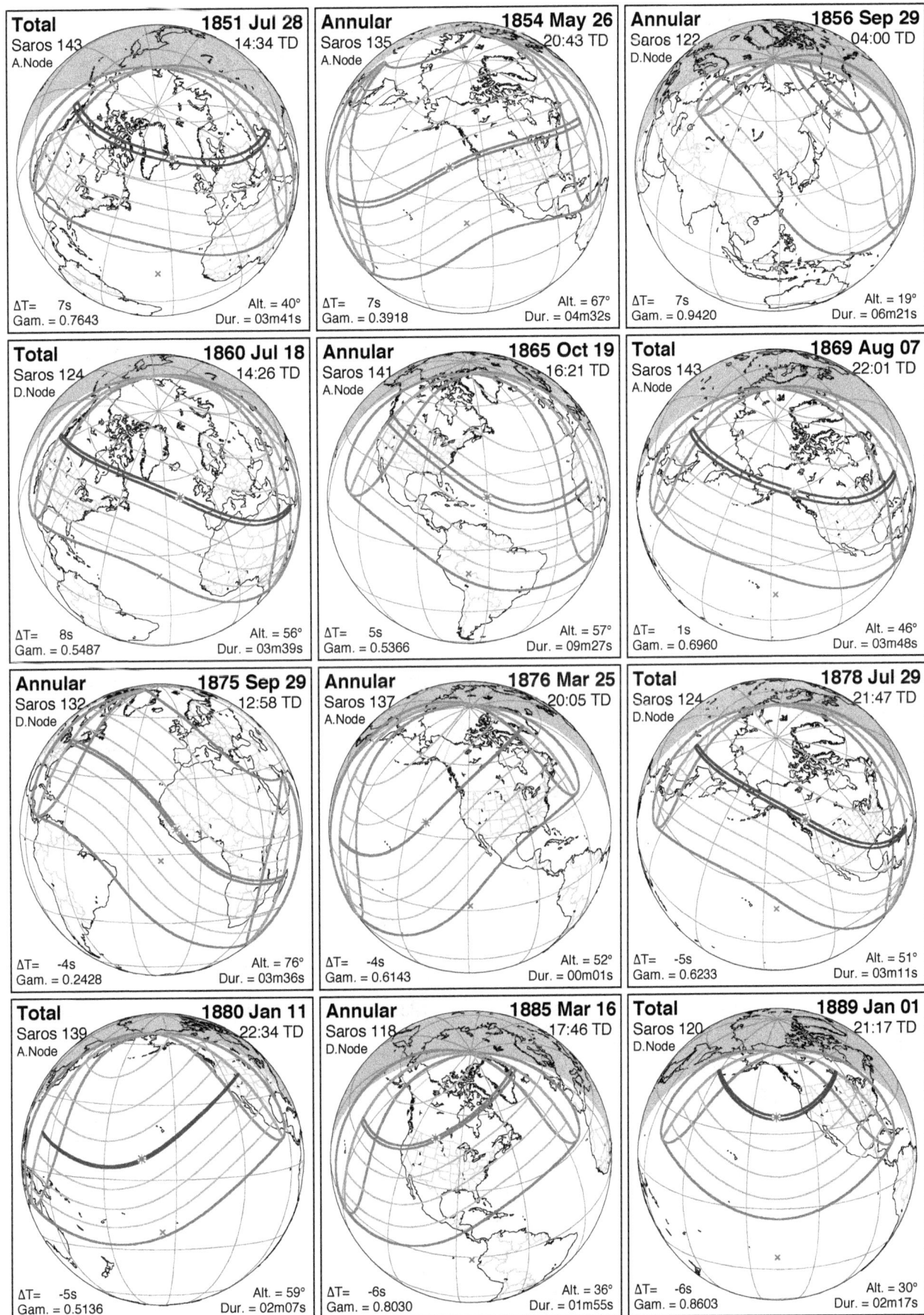

Total	1851 Jul 28
Saros 143	14:34 TD
A.Node	
ΔT= 7s	Alt. = 40°
Gam. = 0.7643	Dur. = 03m41s

Annular	1854 May 26
Saros 135	20:43 TD
A.Node	
ΔT= 7s	Alt. = 67°
Gam. = 0.3918	Dur. = 04m32s

Annular	1856 Sep 29
Saros 122	04:00 TD
D.Node	
ΔT= 7s	Alt. = 19°
Gam. = 0.9420	Dur. = 06m21s

Total	1860 Jul 18
Saros 124	14:26 TD
D.Node	
ΔT= 8s	Alt. = 56°
Gam. = 0.5487	Dur. = 03m39s

Annular	1865 Oct 19
Saros 141	16:21 TD
A.Node	
ΔT= 5s	Alt. = 57°
Gam. = 0.5366	Dur. = 09m27s

Total	1869 Aug 07
Saros 143	22:01 TD
A.Node	
ΔT= 1s	Alt. = 46°
Gam. = 0.6960	Dur. = 03m48s

Annular	1875 Sep 29
Saros 132	12:58 TD
D.Node	
ΔT= -4s	Alt. = 76°
Gam. = 0.2428	Dur. = 03m36s

Annular	1876 Mar 25
Saros 137	20:05 TD
A.Node	
ΔT= -4s	Alt. = 52°
Gam. = 0.6143	Dur. = 00m01s

Total	1878 Jul 29
Saros 124	21:47 TD
D.Node	
ΔT= -5s	Alt. = 51°
Gam. = 0.6233	Dur. = 03m11s

Total	1880 Jan 11
Saros 139	22:34 TD
A.Node	
ΔT= -5s	Alt. = 59°
Gam. = 0.5136	Dur. = 02m07s

Annular	1885 Mar 16
Saros 118	17:46 TD
D.Node	
ΔT= -6s	Alt. = 36°
Gam. = 0.8030	Dur. = 01m55s

Total	1889 Jan 01
Saros 120	21:17 TD
D.Node	
ΔT= -6s	Alt. = 30°
Gam. = 0.8603	Dur. = 02m17s

Plate 019

57

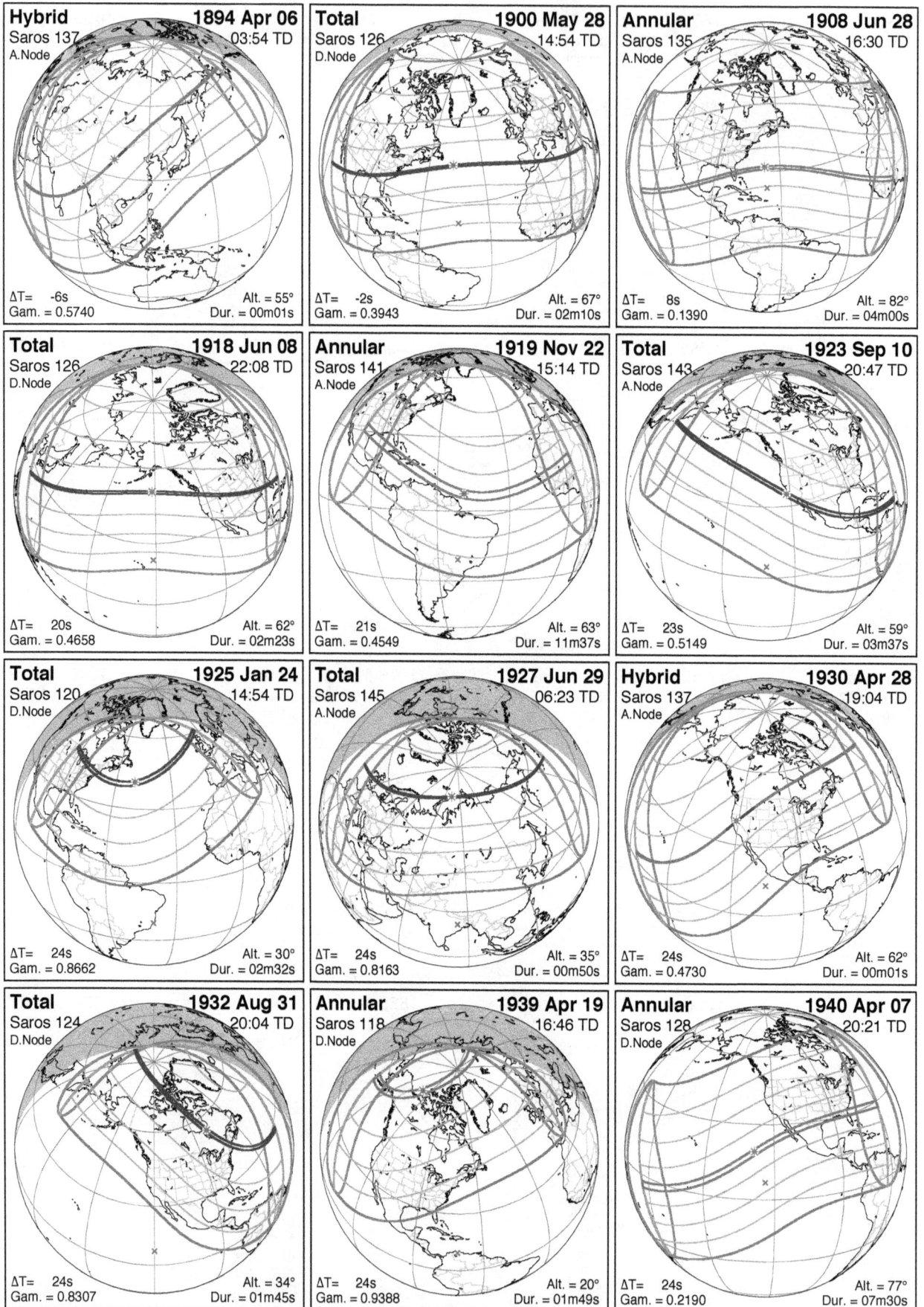

Hybrid	1894 Apr 06
Saros 137	03:54 TD
A.Node	

ΔT= -6s   Alt. = 55°
Gam. = 0.5740   Dur. = 00m01s

Total	1900 May 28
Saros 126	14:54 TD
D.Node	

ΔT= -2s   Alt. = 67°
Gam. = 0.3943   Dur. = 02m10s

Annular	1908 Jun 28
Saros 135	16:30 TD
A.Node	

ΔT= 8s   Alt. = 82°
Gam. = 0.1390   Dur. = 04m00s

Total	1918 Jun 08
Saros 126	22:08 TD
D.Node	

ΔT= 20s   Alt. = 62°
Gam. = 0.4658   Dur. = 02m23s

Annular	1919 Nov 22
Saros 141	15:14 TD
A.Node	

ΔT= 21s   Alt. = 63°
Gam. = 0.4549   Dur. = 11m37s

Total	1923 Sep 10
Saros 143	20:47 TD
A.Node	

ΔT= 23s   Alt. = 59°
Gam. = 0.5149   Dur. = 03m37s

Total	1925 Jan 24
Saros 120	14:54 TD
D.Node	

ΔT= 24s   Alt. = 30°
Gam. = 0.8662   Dur. = 02m32s

Total	1927 Jun 29
Saros 145	06:23 TD
A.Node	

ΔT= 24s   Alt. = 35°
Gam. = 0.8163   Dur. = 00m50s

Hybrid	1930 Apr 28
Saros 137	19:04 TD
A.Node	

ΔT= 24s   Alt. = 62°
Gam. = 0.4730   Dur. = 00m01s

Total	1932 Aug 31
Saros 124	20:04 TD
D.Node	

ΔT= 24s   Alt. = 34°
Gam. = 0.8307   Dur. = 01m45s

Annular	1939 Apr 19
Saros 118	16:46 TD
D.Node	

ΔT= 24s   Alt. = 20°
Gam. = 0.9388   Dur. = 01m49s

Annular	1940 Apr 07
Saros 128	20:21 TD
D.Node	

ΔT= 24s   Alt. = 77°
Gam. = 0.2190   Dur. = 07m30s

Plate 020

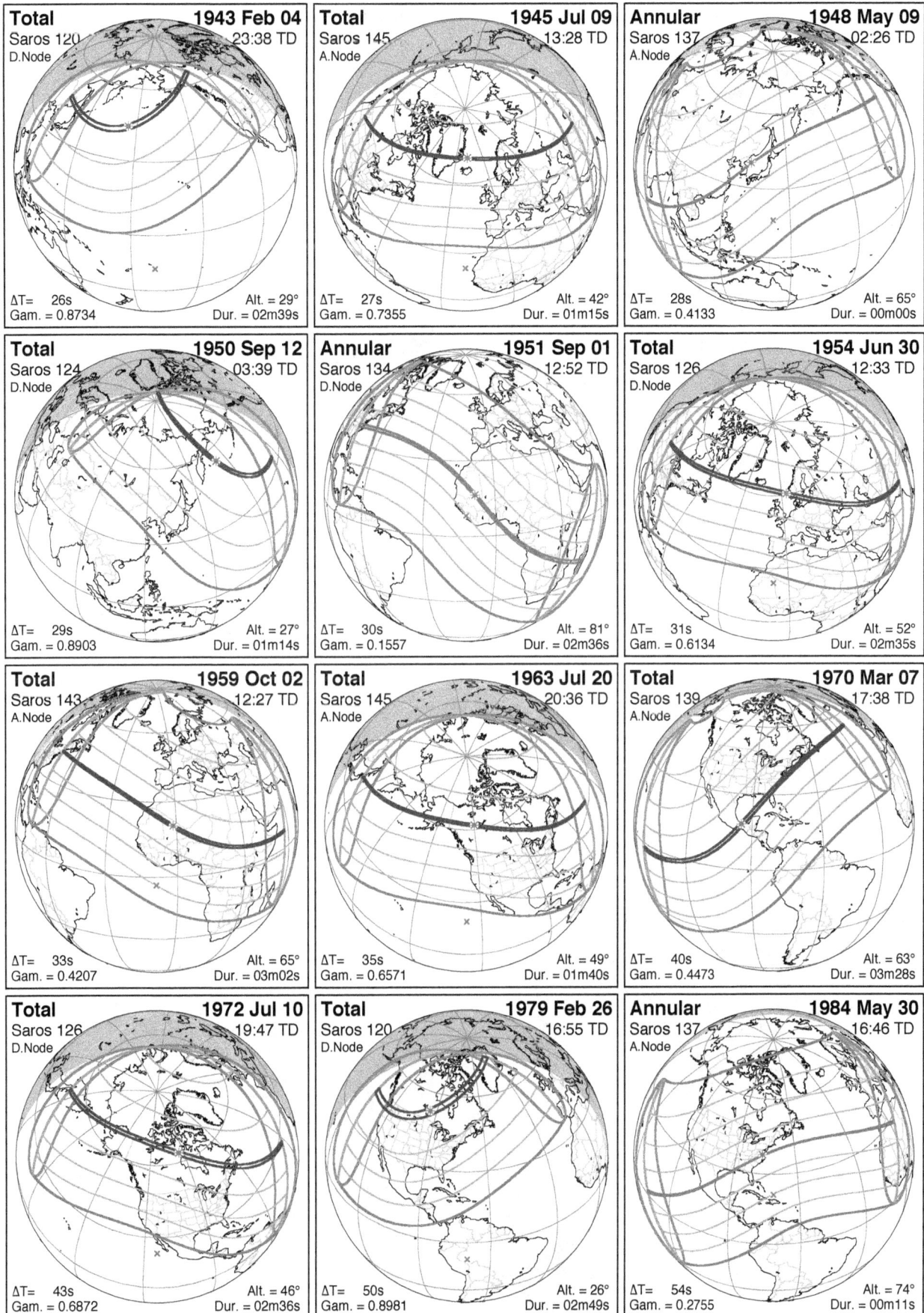

Total	1943 Feb 04
Saros 120	23:38 TD
D.Node	
ΔT= 26s	Alt. = 29°
Gam. = 0.8734	Dur. = 02m39s

Total	1945 Jul 09
Saros 145	13:28 TD
A.Node	
ΔT= 27s	Alt. = 42°
Gam. = 0.7355	Dur. = 01m15s

Annular	1948 May 09
Saros 137	02:26 TD
A.Node	
ΔT= 28s	Alt. = 65°
Gam. = 0.4133	Dur. = 00m00s

Total	1950 Sep 12
Saros 124	03:39 TD
D.Node	
ΔT= 29s	Alt. = 27°
Gam. = 0.8903	Dur. = 01m14s

Annular	1951 Sep 01
Saros 134	12:52 TD
D.Node	
ΔT= 30s	Alt. = 81°
Gam. = 0.1557	Dur. = 02m36s

Total	1954 Jun 30
Saros 126	12:33 TD
D.Node	
ΔT= 31s	Alt. = 52°
Gam. = 0.6134	Dur. = 02m35s

Total	1959 Oct 02
Saros 143	12:27 TD
A.Node	
ΔT= 33s	Alt. = 65°
Gam. = 0.4207	Dur. = 03m02s

Total	1963 Jul 20
Saros 145	20:36 TD
A.Node	
ΔT= 35s	Alt. = 49°
Gam. = 0.6571	Dur. = 01m40s

Total	1970 Mar 07
Saros 139	17:38 TD
A.Node	
ΔT= 40s	Alt. = 63°
Gam. = 0.4473	Dur. = 03m28s

Total	1972 Jul 10
Saros 126	19:47 TD
D.Node	
ΔT= 43s	Alt. = 46°
Gam. = 0.6872	Dur. = 02m36s

Total	1979 Feb 26
Saros 120	16:55 TD
D.Node	
ΔT= 50s	Alt. = 26°
Gam. = 0.8981	Dur. = 02m49s

Annular	1984 May 30
Saros 137	16:46 TD
A.Node	
ΔT= 54s	Alt. = 74°
Gam. = 0.2755	Dur. = 00m11s

Plate 021

59

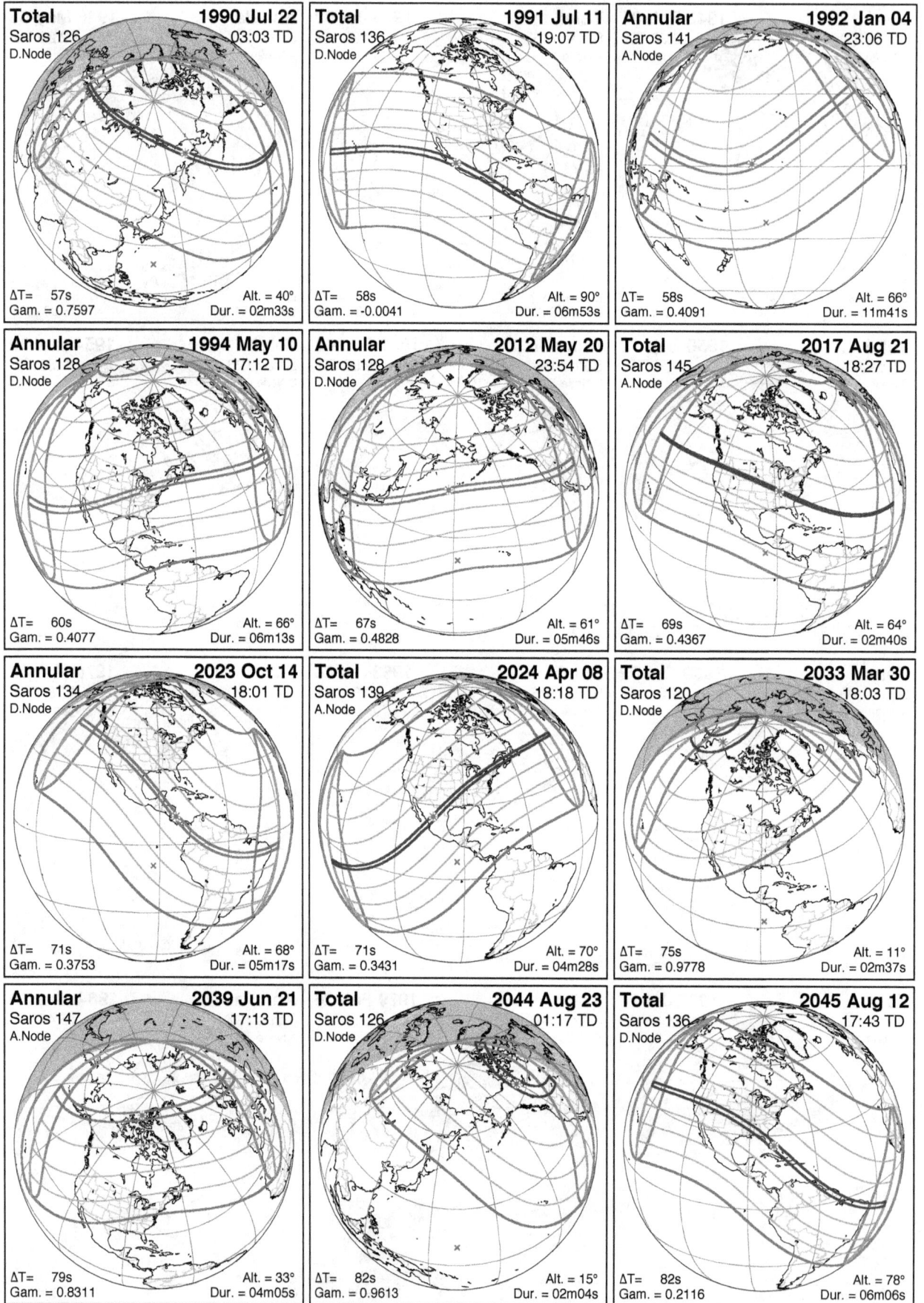

Total **1990 Jul 22**	
Saros 126 03:03 TD	
D.Node	
ΔT= 57s Alt. = 40°	
Gam. = 0.7597 Dur. = 02m33s	

Total — **1990 Jul 22** — Saros 126 — 03:03 TD — D.Node — ΔT= 57s — Gam. = 0.7597 — Alt. = 40° — Dur. = 02m33s

Total — **1991 Jul 11** — Saros 136 — 19:07 TD — D.Node — ΔT= 58s — Gam. = -0.0041 — Alt. = 90° — Dur. = 06m53s

Annular — **1992 Jan 04** — Saros 141 — 23:06 TD — A.Node — ΔT= 58s — Gam. = 0.4091 — Alt. = 66° — Dur. = 11m41s

Annular — **1994 May 10** — Saros 128 — 17:12 TD — D.Node — ΔT= 60s — Gam. = 0.4077 — Alt. = 66° — Dur. = 06m13s

Annular — **2012 May 20** — Saros 128 — 23:54 TD — D.Node — ΔT= 67s — Gam. = 0.4828 — Alt. = 61° — Dur. = 05m46s

Total — **2017 Aug 21** — Saros 145 — 18:27 TD — A.Node — ΔT= 69s — Gam. = 0.4367 — Alt. = 64° — Dur. = 02m40s

Annular — **2023 Oct 14** — Saros 134 — 18:01 TD — D.Node — ΔT= 71s — Gam. = 0.3753 — Alt. = 68° — Dur. = 05m17s

Total — **2024 Apr 08** — Saros 139 — 18:18 TD — A.Node — ΔT= 71s — Gam. = 0.3431 — Alt. = 70° — Dur. = 04m28s

Total — **2033 Mar 30** — Saros 120 — 18:03 TD — D.Node — ΔT= 75s — Gam. = 0.9778 — Alt. = 11° — Dur. = 02m37s

Annular — **2039 Jun 21** — Saros 147 — 17:13 TD — A.Node — ΔT= 79s — Gam. = 0.8311 — Alt. = 33° — Dur. = 04m05s

Total — **2044 Aug 23** — Saros 126 — 01:17 TD — D.Node — ΔT= 82s — Gam. = 0.9613 — Alt. = 15° — Dur. = 02m04s

Total — **2045 Aug 12** — Saros 136 — 17:43 TD — D.Node — ΔT= 82s — Gam. = 0.2116 — Alt. = 78° — Dur. = 06m06s

Plate 022

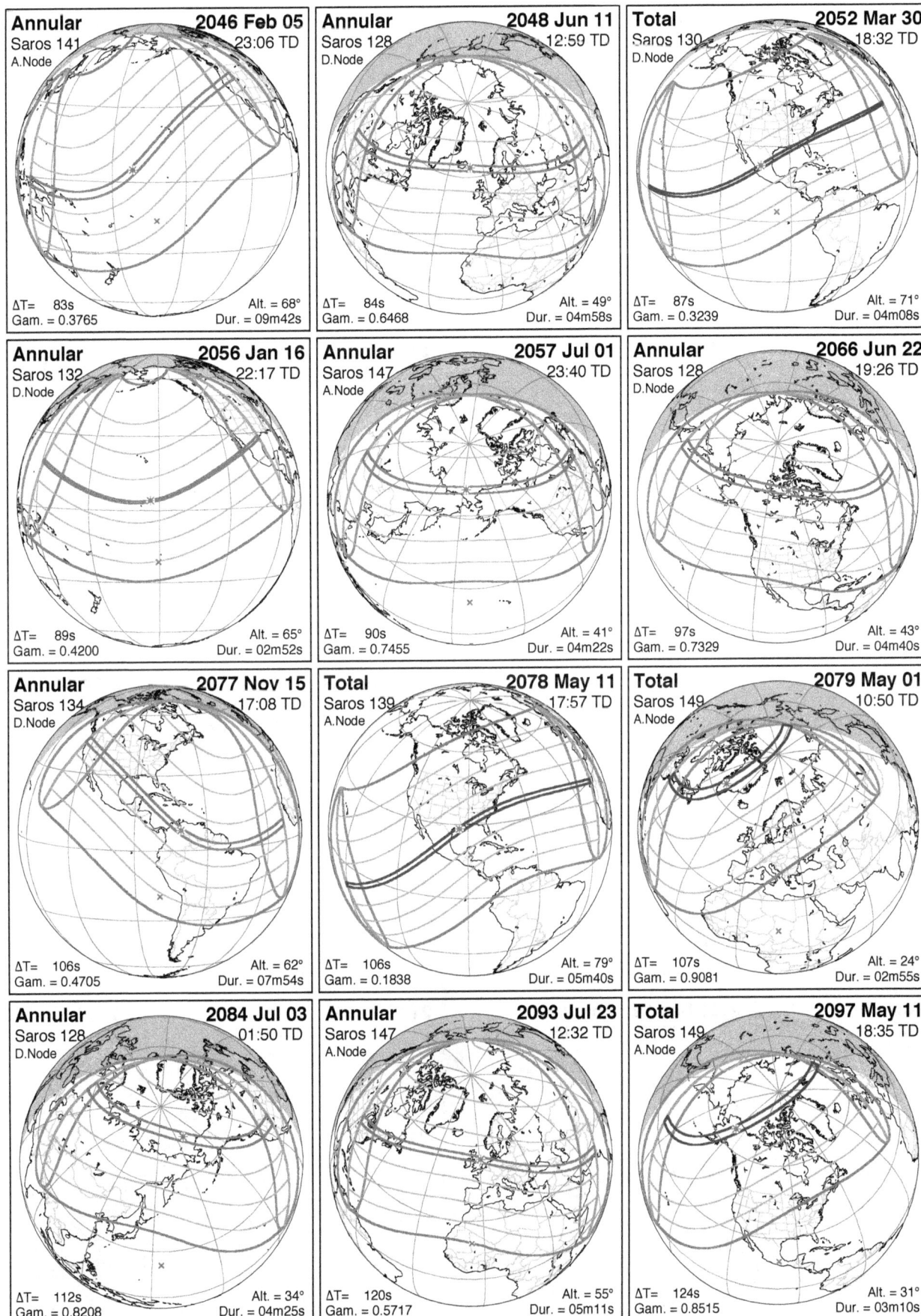

Annular **2046 Feb 05**	
Saros 141 23:06 TD	
A.Node	
ΔT= 83s	Alt. = 68°
Gam. = 0.3765	Dur. = 09m42s

Annular **2048 Jun 11**	
Saros 128 12:59 TD	
D.Node	
ΔT= 84s	Alt. = 49°
Gam. = 0.6468	Dur. = 04m58s

Total **2052 Mar 30**	
Saros 130 18:32 TD	
D.Node	
ΔT= 87s	Alt. = 71°
Gam. = 0.3239	Dur. = 04m08s

Annular **2056 Jan 16**	
Saros 132 22:17 TD	
D.Node	
ΔT= 89s	Alt. = 65°
Gam. = 0.4200	Dur. = 02m52s

Annular **2057 Jul 01**	
Saros 147 23:40 TD	
A.Node	
ΔT= 90s	Alt. = 41°
Gam. = 0.7455	Dur. = 04m22s

Annular **2066 Jun 22**	
Saros 128 19:26 TD	
D.Node	
ΔT= 97s	Alt. = 43°
Gam. = 0.7329	Dur. = 04m40s

Annular **2077 Nov 15**	
Saros 134 17:08 TD	
D.Node	
ΔT= 106s	Alt. = 62°
Gam. = 0.4705	Dur. = 07m54s

Total **2078 May 11**	
Saros 139 17:57 TD	
A.Node	
ΔT= 106s	Alt. = 79°
Gam. = 0.1838	Dur. = 05m40s

Total **2079 May 01**	
Saros 149 10:50 TD	
A.Node	
ΔT= 107s	Alt. = 24°
Gam. = 0.9081	Dur. = 02m55s

Annular **2084 Jul 03**	
Saros 128 01:50 TD	
D.Node	
ΔT= 112s	Alt. = 34°
Gam. = 0.8208	Dur. = 04m25s

Annular **2093 Jul 23**	
Saros 147 12:32 TD	
A.Node	
ΔT= 120s	Alt. = 55°
Gam. = 0.5717	Dur. = 05m11s

Total **2097 May 11**	
Saros 149 18:35 TD	
A.Node	
ΔT= 124s	Alt. = 31°
Gam. = 0.8515	Dur. = 03m10s

Plate 023

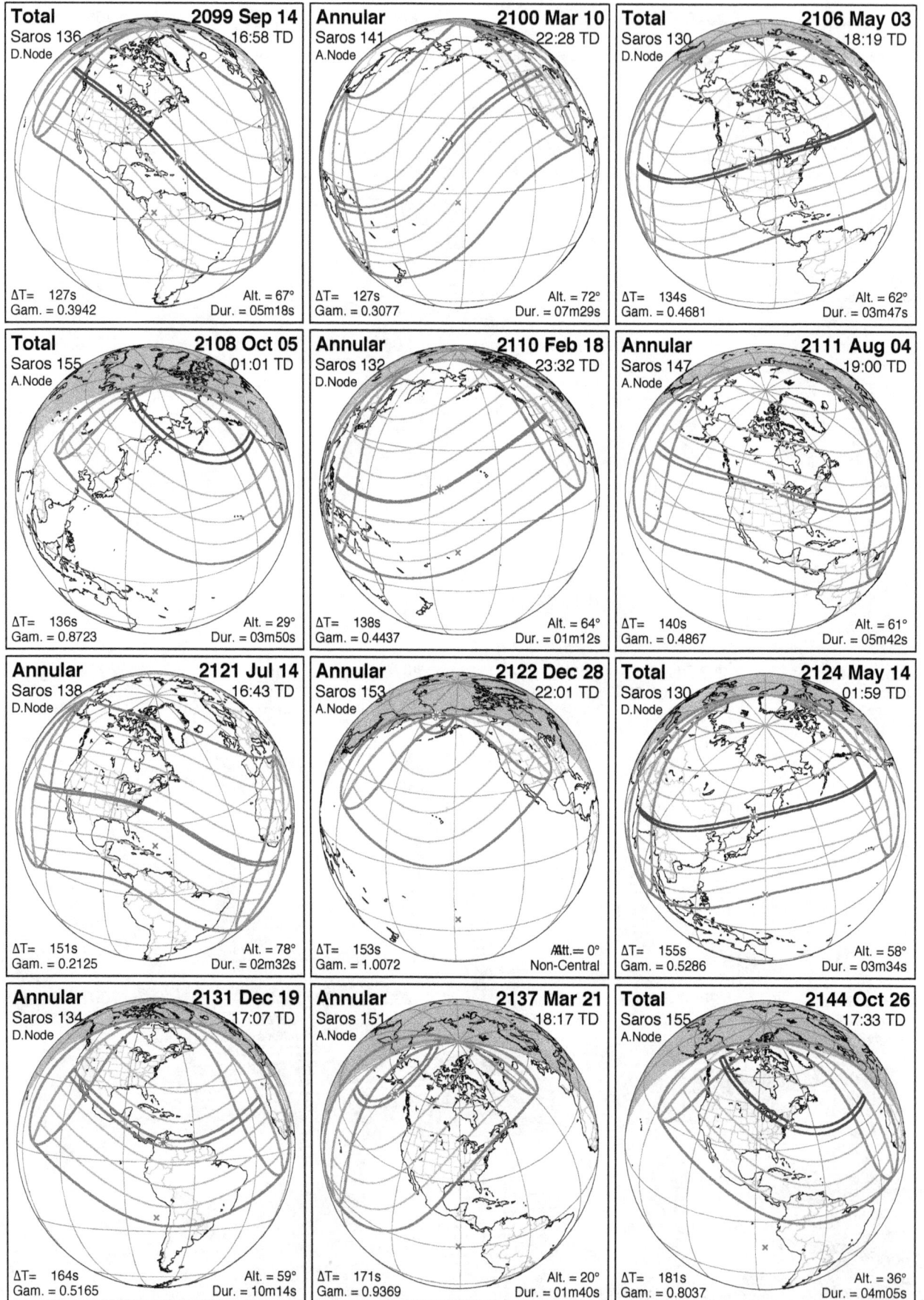

Total	2099 Sep 14
Saros 136	16:58 TD
D.Node	
ΔT= 127s	Alt. = 67°
Gam. = 0.3942	Dur. = 05m18s

Annular	2100 Mar 10
Saros 141	22:28 TD
A.Node	
ΔT= 127s	Alt. = 72°
Gam. = 0.3077	Dur. = 07m29s

Total	2106 May 03
Saros 130	18:19 TD
D.Node	
ΔT= 134s	Alt. = 62°
Gam. = 0.4681	Dur. = 03m47s

Total	2108 Oct 05
Saros 155	01:01 TD
A.Node	
ΔT= 136s	Alt. = 29°
Gam. = 0.8723	Dur. = 03m50s

Annular	2110 Feb 18
Saros 132	23:32 TD
D.Node	
ΔT= 138s	Alt. = 64°
Gam. = 0.4437	Dur. = 01m12s

Annular	2111 Aug 04
Saros 147	19:00 TD
A.Node	
ΔT= 140s	Alt. = 61°
Gam. = 0.4867	Dur. = 05m42s

Annular	2121 Jul 14
Saros 138	16:43 TD
D.Node	
ΔT= 151s	Alt. = 78°
Gam. = 0.2125	Dur. = 02m32s

Annular	2122 Dec 28
Saros 153	22:01 TD
A.Node	
ΔT= 153s	Alt.= 0°
Gam. = 1.0072	Non-Central

Total	2124 May 14
Saros 130	01:59 TD
D.Node	
ΔT= 155s	Alt. = 58°
Gam. = 0.5286	Dur. = 03m34s

Annular	2131 Dec 19
Saros 134	17:07 TD
D.Node	
ΔT= 164s	Alt. = 59°
Gam. = 0.5165	Dur. = 10m14s

Annular	2137 Mar 21
Saros 151	18:17 TD
A.Node	
ΔT= 171s	Alt. = 20°
Gam. = 0.9369	Dur. = 01m40s

Total	2144 Oct 26
Saros 155	17:33 TD
A.Node	
ΔT= 181s	Alt. = 36°
Gam. = 0.8037	Dur. = 04m05s

Plate 024

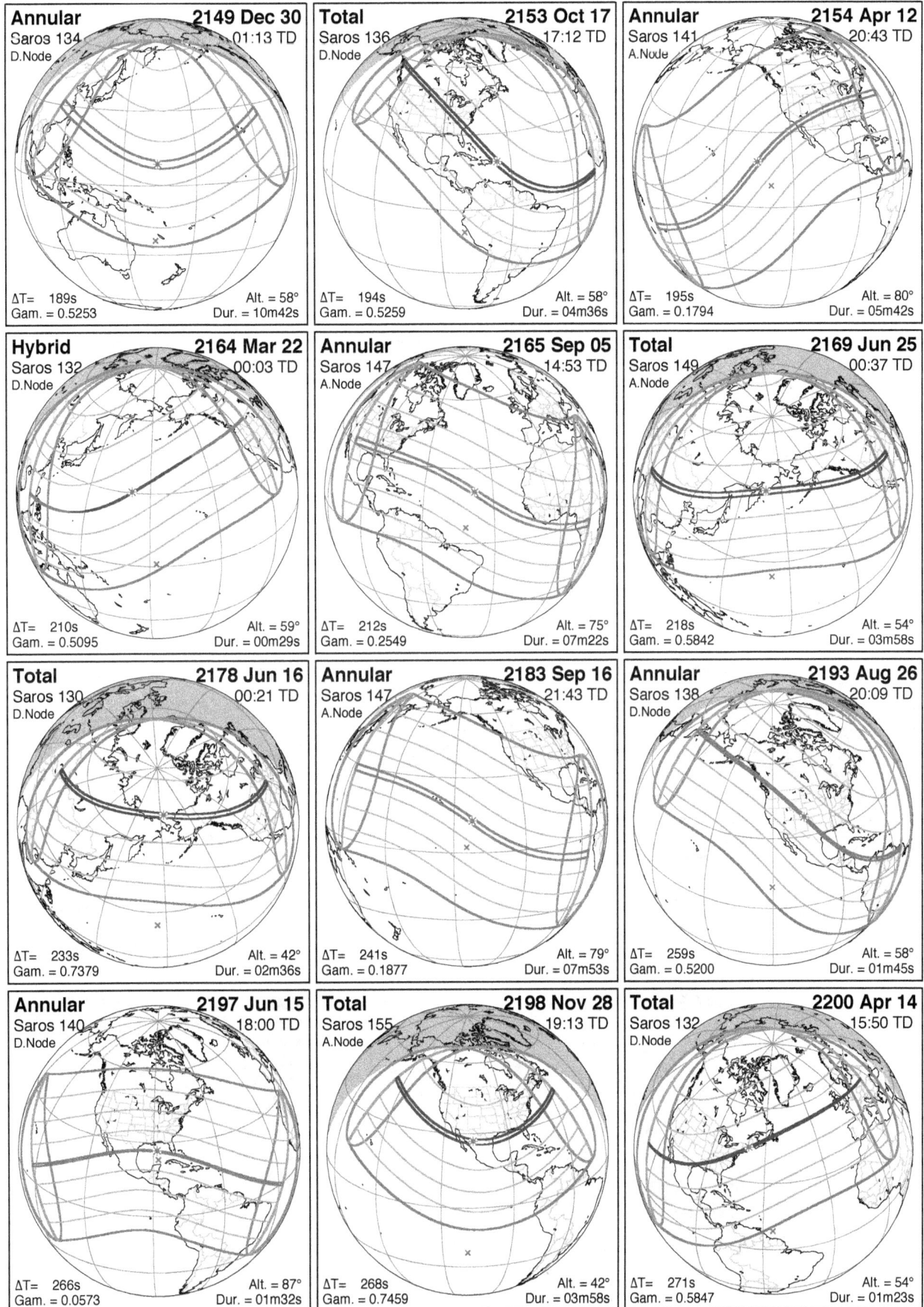

Annular	2149 Dec 30
Saros 134	01:13 TD
D.Node	
ΔT= 189s	Alt. = 58°
Gam. = 0.5253	Dur. = 10m42s

Total	2153 Oct 17
Saros 136	17:12 TD
D.Node	
ΔT= 194s	Alt. = 58°
Gam. = 0.5259	Dur. = 04m36s

Annular	2154 Apr 12
Saros 141	20:43 TD
A.Node	
ΔT= 195s	Alt. = 80°
Gam. = 0.1794	Dur. = 05m42s

Hybrid	2164 Mar 22
Saros 132	00:03 TD
D.Node	
ΔT= 210s	Alt. = 59°
Gam. = 0.5095	Dur. = 00m29s

Annular	2165 Sep 05
Saros 147	14:53 TD
A.Node	
ΔT= 212s	Alt. = 75°
Gam. = 0.2549	Dur. = 07m22s

Total	2169 Jun 25
Saros 149	00:37 TD
A.Node	
ΔT= 218s	Alt. = 54°
Gam. = 0.5842	Dur. = 03m58s

Total	2178 Jun 16
Saros 130	00:21 TD
D.Node	
ΔT= 233s	Alt. = 42°
Gam. = 0.7379	Dur. = 02m36s

Annular	2183 Sep 16
Saros 147	21:43 TD
A.Node	
ΔT= 241s	Alt. = 79°
Gam. = 0.1877	Dur. = 07m53s

Annular	2193 Aug 26
Saros 138	20:09 TD
D.Node	
ΔT= 259s	Alt. = 58°
Gam. = 0.5200	Dur. = 01m45s

Annular	2197 Jun 15
Saros 140	18:00 TD
D.Node	
ΔT= 266s	Alt. = 87°
Gam. = 0.0573	Dur. = 01m32s

Total	2198 Nov 28
Saros 155	19:13 TD
A.Node	
ΔT= 268s	Alt. = 42°
Gam. = 0.7459	Dur. = 03m58s

Total	2200 Apr 14
Saros 132	15:50 TD
D.Node	
ΔT= 271s	Alt. = 54°
Gam. = 0.5847	Dur. = 01m23s

Plate 025

63

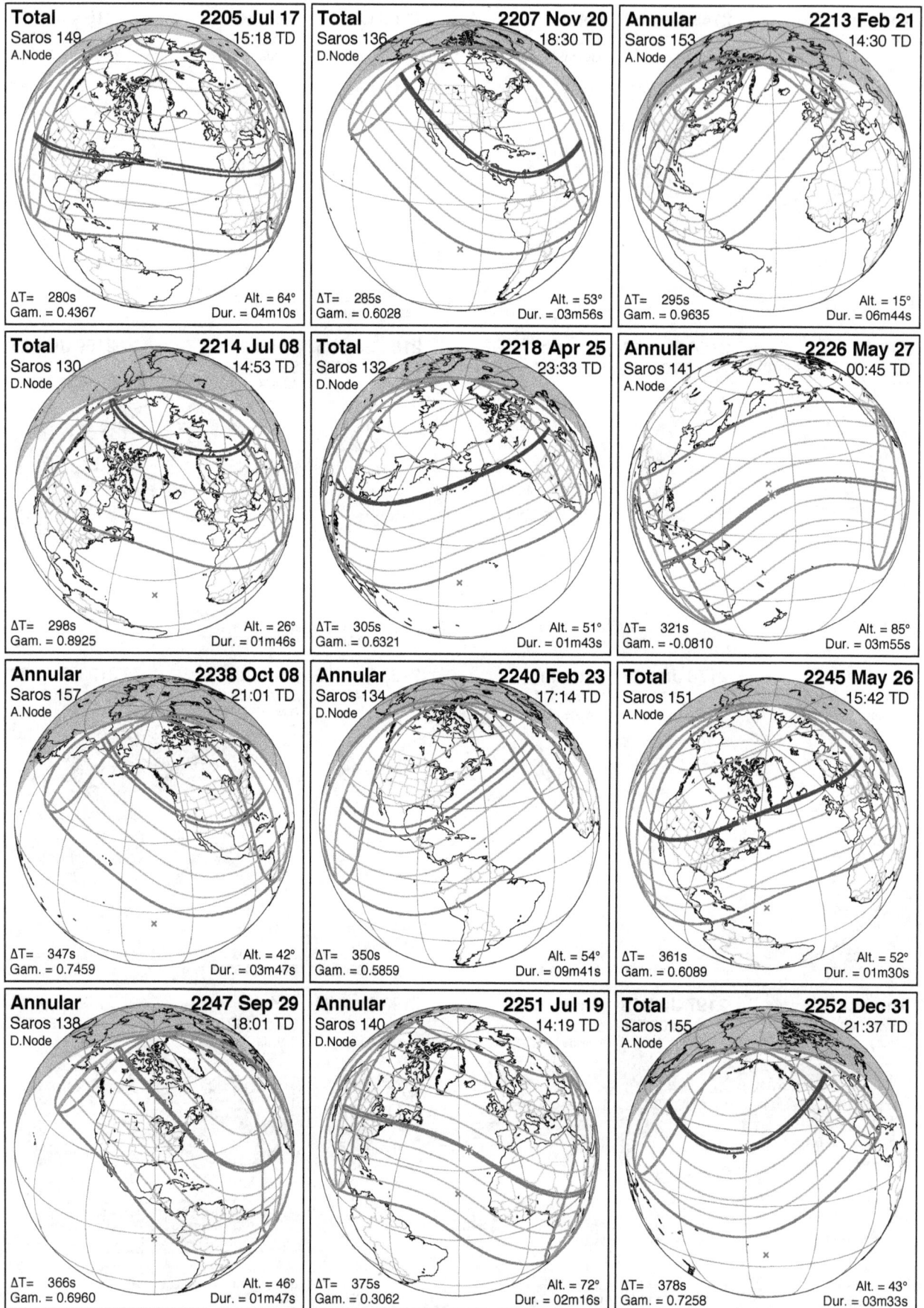

Total **2205 Jul 17** Saros 149 15:18 TD A.Node	

Total — **2205 Jul 17**
Saros 149 — 15:18 TD
A.Node
ΔT= 280s — Alt. = 64°
Gam. = 0.4367 — Dur. = 04m10s

Total — **2207 Nov 20**
Saros 136 — 18:30 TD
D.Node
ΔT= 285s — Alt. = 53°
Gam. = 0.6028 — Dur. = 03m56s

Annular — **2213 Feb 21**
Saros 153 — 14:30 TD
A.Node
ΔT= 295s — Alt. = 15°
Gam. = 0.9635 — Dur. = 06m44s

Total — **2214 Jul 08**
Saros 130 — 14:53 TD
D.Node
ΔT= 298s — Alt. = 26°
Gam. = 0.8925 — Dur. = 01m46s

Total — **2218 Apr 25**
Saros 132 — 23:33 TD
D.Node
ΔT= 305s — Alt. = 51°
Gam. = 0.6321 — Dur. = 01m43s

Annular — **2226 May 27**
Saros 141 — 00:45 TD
A.Node
ΔT= 321s — Alt. = 85°
Gam. = -0.0810 — Dur. = 03m55s

Annular — **2238 Oct 08**
Saros 157 — 21:01 TD
A.Node
ΔT= 347s — Alt. = 42°
Gam. = 0.7459 — Dur. = 03m47s

Annular — **2240 Feb 23**
Saros 134 — 17:14 TD
D.Node
ΔT= 350s — Alt. = 54°
Gam. = 0.5859 — Dur. = 09m41s

Total — **2245 May 26**
Saros 151 — 15:42 TD
A.Node
ΔT= 361s — Alt. = 52°
Gam. = 0.6089 — Dur. = 01m30s

Annular — **2247 Sep 29**
Saros 138 — 18:01 TD
D.Node
ΔT= 366s — Alt. = 46°
Gam. = 0.6960 — Dur. = 01m47s

Annular — **2251 Jul 19**
Saros 140 — 14:19 TD
D.Node
ΔT= 375s — Alt. = 72°
Gam. = 0.3062 — Dur. = 02m16s

Total — **2252 Dec 31**
Saros 155 — 21:37 TD
A.Node
ΔT= 378s — Alt. = 43°
Gam. = 0.7258 — Dur. = 03m33s

Plate 026

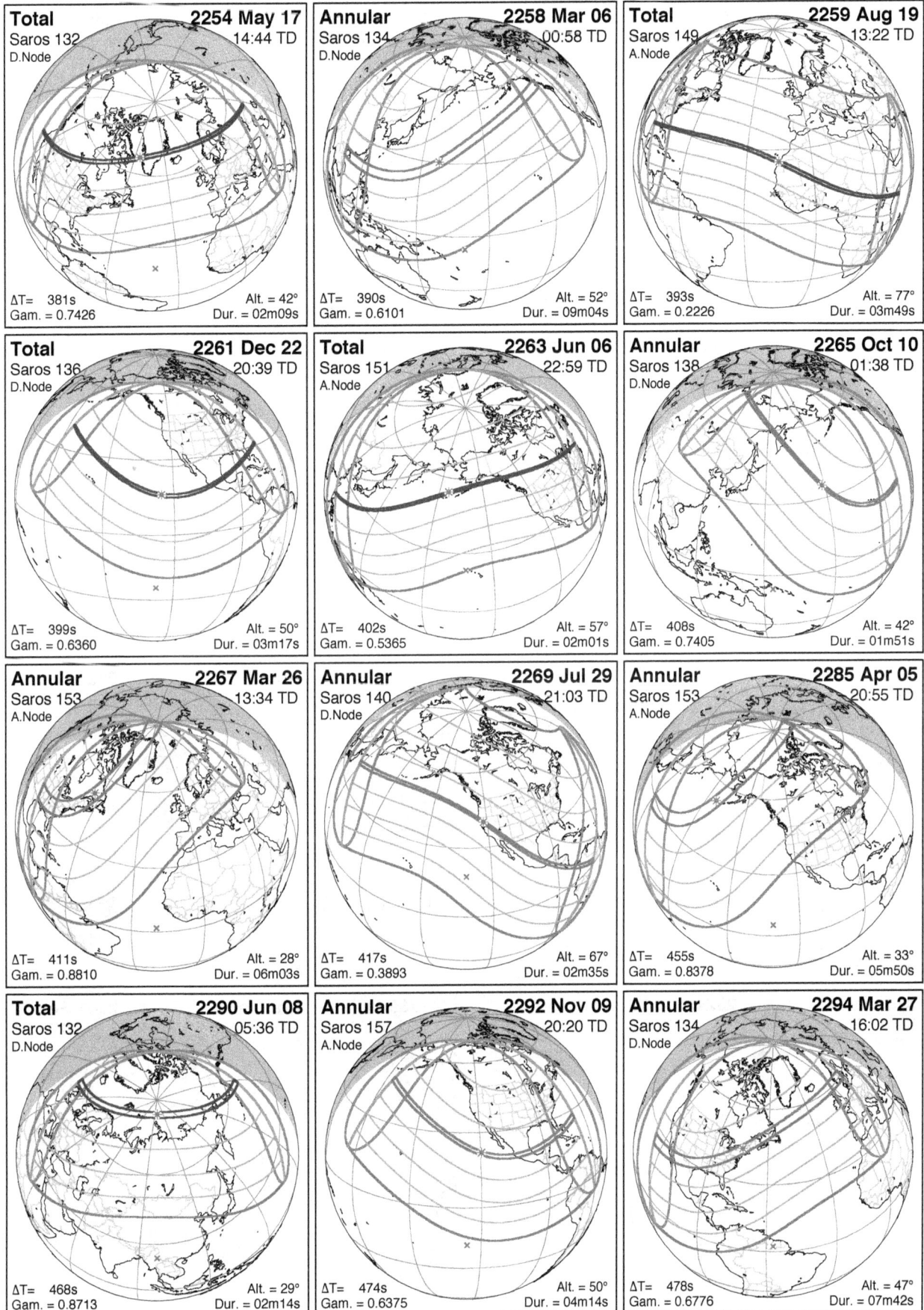

Total	2254 May 17
Saros 132	14:44 TD
D.Node	
ΔT= 381s	Alt. = 42°
Gam. = 0.7426	Dur. = 02m09s

Annular	2258 Mar 06
Saros 134	00:58 TD
D.Node	
ΔT= 390s	Alt. = 52°
Gam. = 0.6101	Dur. = 09m04s

Total	2259 Aug 19
Saros 149	13:22 TD
A.Node	
ΔT= 393s	Alt. = 77°
Gam. = 0.2226	Dur. = 03m49s

Total	2261 Dec 22
Saros 136	20:39 TD
D.Node	
ΔT= 399s	Alt. = 50°
Gam. = 0.6360	Dur. = 03m17s

Total	2263 Jun 06
Saros 151	22:59 TD
A.Node	
ΔT= 402s	Alt. = 57°
Gam. = 0.5365	Dur. = 02m01s

Annular	2265 Oct 10
Saros 138	01:38 TD
D.Node	
ΔT= 408s	Alt. = 42°
Gam. = 0.7405	Dur. = 01m51s

Annular	2267 Mar 26
Saros 153	13:34 TD
A.Node	
ΔT= 411s	Alt. = 28°
Gam. = 0.8810	Dur. = 06m03s

Annular	2269 Jul 29
Saros 140	21:03 TD
D.Node	
ΔT= 417s	Alt. = 67°
Gam. = 0.3893	Dur. = 02m35s

Annular	2285 Apr 05
Saros 153	20:55 TD
A.Node	
ΔT= 455s	Alt. = 33°
Gam. = 0.8378	Dur. = 05m50s

Total	2290 Jun 08
Saros 132	05:36 TD
D.Node	
ΔT= 468s	Alt. = 29°
Gam. = 0.8713	Dur. = 02m14s

Annular	2292 Nov 09
Saros 157	20:20 TD
A.Node	
ΔT= 474s	Alt. = 50°
Gam. = 0.6375	Dur. = 04m14s

Annular	2294 Mar 27
Saros 134	16:02 TD
D.Node	
ΔT= 478s	Alt. = 47°
Gam. = 0.6776	Dur. = 07m42s

Plate 027

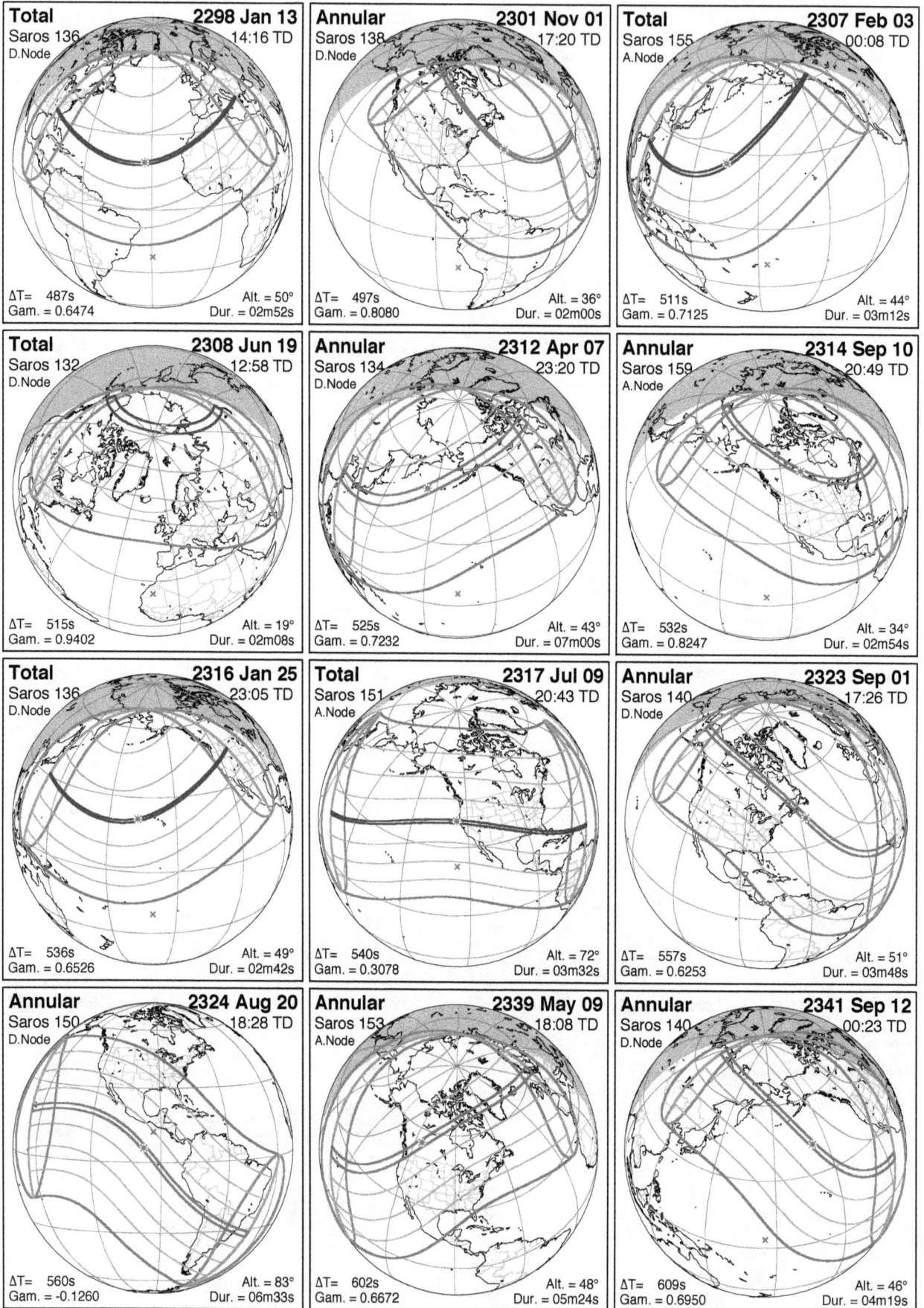

Total	2298 Jan 13
Saros 136	14:16 TD
D.Node	
ΔT= 487s	Alt. = 50°
Gam. = 0.6474	Dur. = 02m52s

Annular	2301 Nov 01
Saros 138	17:20 TD
D.Node	
ΔT= 497s	Alt. = 36°
Gam. = 0.8080	Dur. = 02m00s

Total	2307 Feb 03
Saros 155	00:08 TD
A.Node	
ΔT= 511s	Alt. = 44°
Gam. = 0.7125	Dur. = 03m12s

Total	2308 Jun 19
Saros 132	12:58 TD
D.Node	
ΔT= 515s	Alt. = 19°
Gam. = 0.9402	Dur. = 02m08s

Annular	2312 Apr 07
Saros 134	23:20 TD
D.Node	
ΔT= 525s	Alt. = 43°
Gam. = 0.7232	Dur. = 07m00s

Annular	2314 Sep 10
Saros 159	20:49 TD
A.Node	
ΔT= 532s	Alt. = 34°
Gam. = 0.8247	Dur. = 02m54s

Total	2316 Jan 25
Saros 136	23:05 TD
D.Node	
ΔT= 536s	Alt. = 49°
Gam. = 0.6526	Dur. = 02m42s

Total	2317 Jul 09
Saros 151	20:43 TD
A.Node	
ΔT= 540s	Alt. = 72°
Gam. = 0.3078	Dur. = 03m32s

Annular	2323 Sep 01
Saros 140	17:26 TD
D.Node	
ΔT= 557s	Alt. = 51°
Gam. = 0.6253	Dur. = 03m48s

Annular	2324 Aug 20
Saros 150	18:28 TD
D.Node	
ΔT= 560s	Alt. = 83°
Gam. = -0.1260	Dur. = 06m33s

Annular	2339 May 09
Saros 153	18:08 TD
A.Node	
ΔT= 602s	Alt. = 48°
Gam. = 0.6672	Dur. = 05m24s

Annular	2341 Sep 12
Saros 140	00:23 TD
D.Node	
ΔT= 609s	Alt. = 46°
Gam. = 0.6950	Dur. = 04m19s

Plate 028

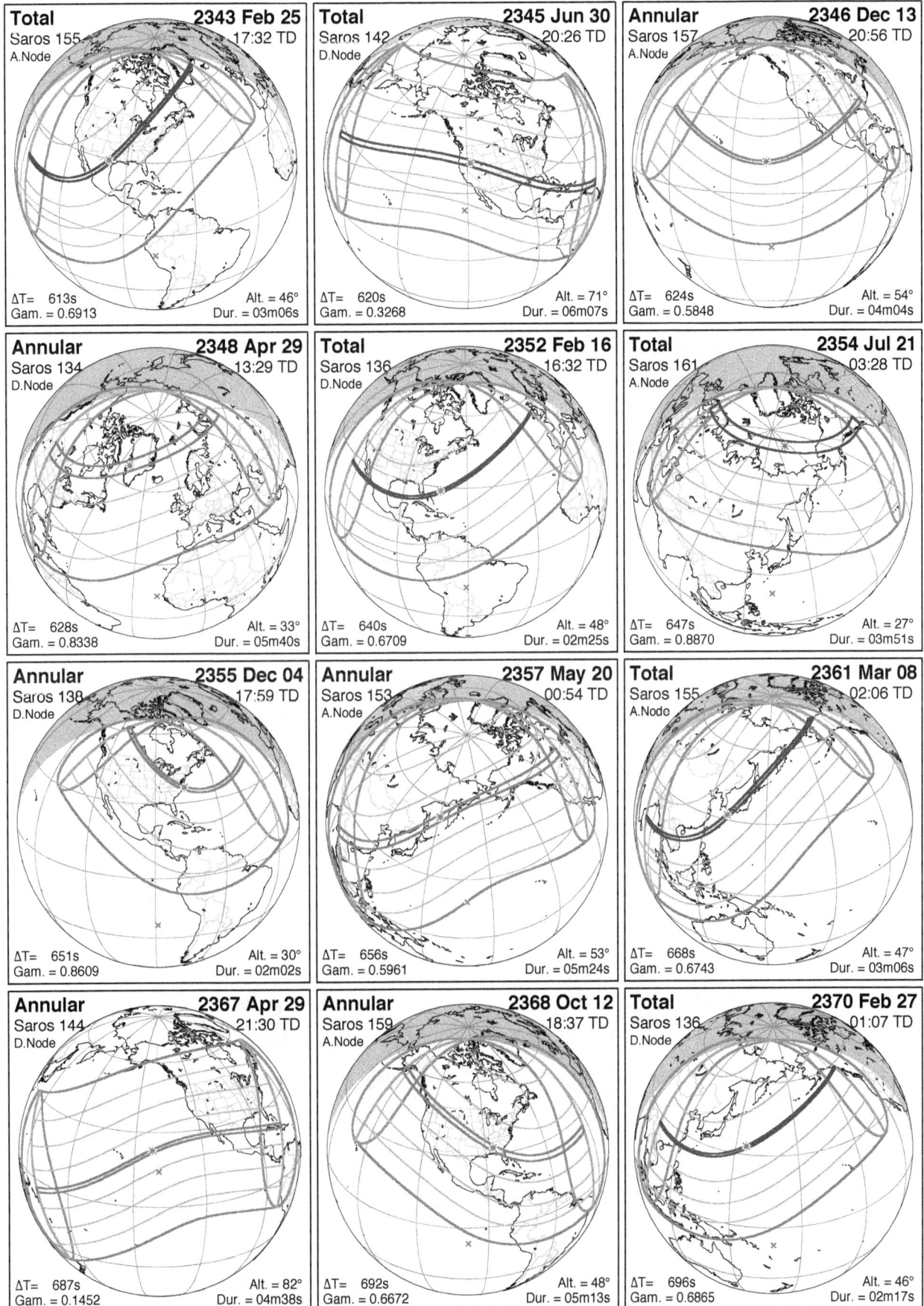

Total	**2343 Feb 25**
Saros 155	17:32 TD
A.Node	
ΔT= 613s	Alt. = 46°
Gam. = 0.6913	Dur. = 03m06s

Total	**2345 Jun 30**
Saros 142	20:26 TD
D.Node	
ΔT= 620s	Alt. = 71°
Gam. = 0.3268	Dur. = 06m07s

Annular	**2346 Dec 13**
Saros 157	20:56 TD
A.Node	
ΔT= 624s	Alt. = 54°
Gam. = 0.5848	Dur. = 04m04s

Annular	**2348 Apr 29**
Saros 134	13:29 TD
D.Node	
ΔT= 628s	Alt. = 33°
Gam. = 0.8338	Dur. = 05m40s

Total	**2352 Feb 16**
Saros 136	16:32 TD
D.Node	
ΔT= 640s	Alt. = 48°
Gam. = 0.6709	Dur. = 02m25s

Total	**2354 Jul 21**
Saros 161	03:28 TD
A.Node	
ΔT= 647s	Alt. = 27°
Gam. = 0.8870	Dur. = 03m51s

Annular	**2355 Dec 04**
Saros 138	17:59 TD
D.Node	
ΔT= 651s	Alt. = 30°
Gam. = 0.8609	Dur. = 02m02s

Annular	**2357 May 20**
Saros 153	00:54 TD
A.Node	
ΔT= 656s	Alt. = 53°
Gam. = 0.5961	Dur. = 05m24s

Total	**2361 Mar 08**
Saros 155	02:06 TD
A.Node	
ΔT= 668s	Alt. = 47°
Gam. = 0.6743	Dur. = 03m06s

Annular	**2367 Apr 29**
Saros 144	21:30 TD
D.Node	
ΔT= 687s	Alt. = 82°
Gam. = 0.1452	Dur. = 04m38s

Annular	**2368 Oct 12**
Saros 159	18:37 TD
A.Node	
ΔT= 692s	Alt. = 48°
Gam. = 0.6672	Dur. = 05m13s

Total	**2370 Feb 27**
Saros 136	01:07 TD
D.Node	
ΔT= 696s	Alt. = 46°
Gam. = 0.6865	Dur. = 02m17s

Plate 029

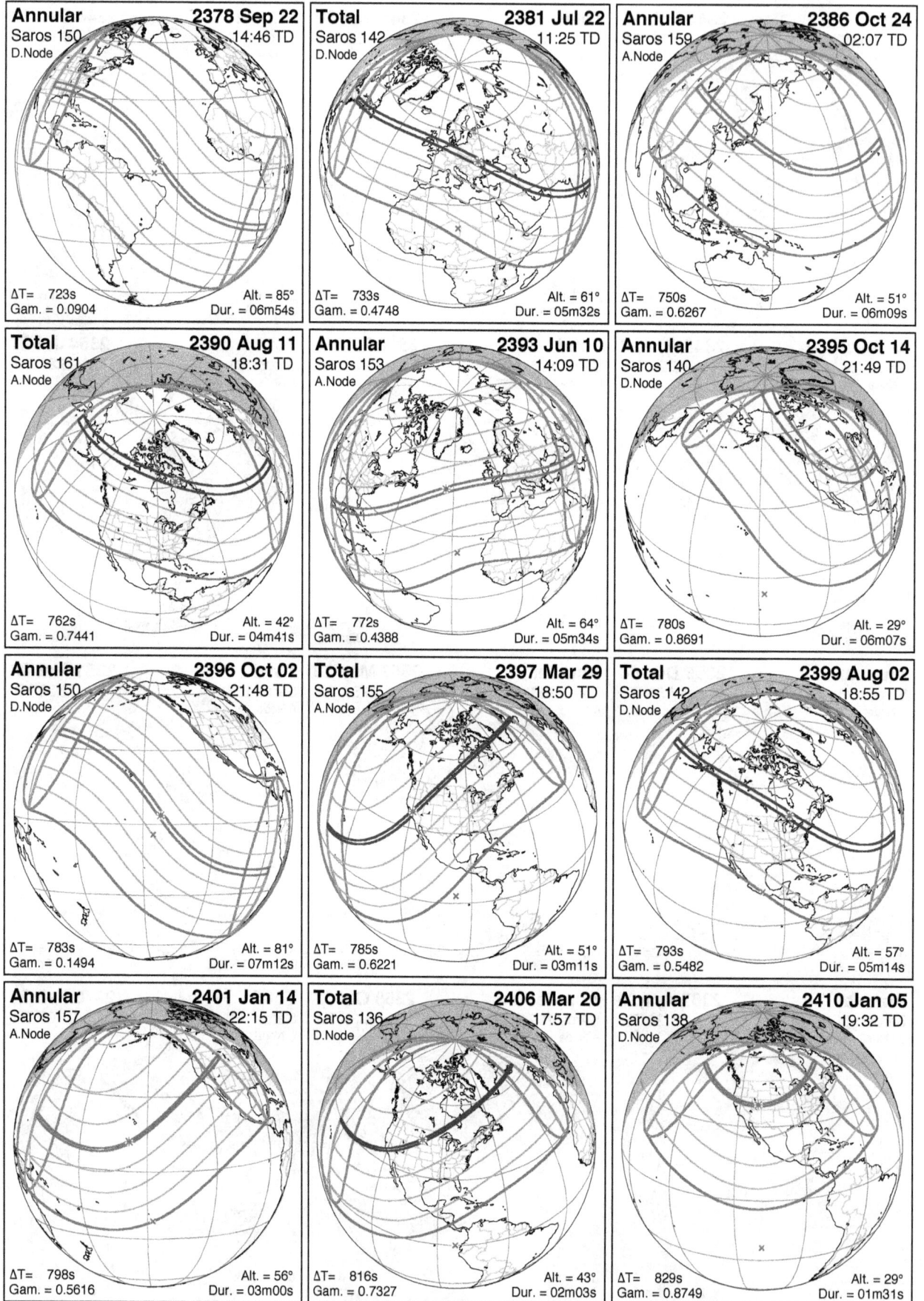

Annular **2378 Sep 22** Saros 150 14:46 TD D.Node ΔT= 723s Alt. = 85° Gam. = 0.0904 Dur. = 06m54s	**Total** **2381 Jul 22** Saros 142 11:25 TD D.Node ΔT= 733s Alt. = 61° Gam. = 0.4748 Dur. = 05m32s
Annular **2386 Oct 24** Saros 159 02:07 TD A.Node ΔT= 750s Alt. = 51° Gam. = 0.6267 Dur. = 06m09s	

Total **2390 Aug 11** Saros 161 18:31 TD A.Node ΔT= 762s Alt. = 42° Gam. = 0.7441 Dur. = 04m41s	**Annular** **2393 Jun 10** Saros 153 14:09 TD A.Node ΔT= 772s Alt. = 64° Gam. = 0.4388 Dur. = 05m34s
Annular **2395 Oct 14** Saros 140 21:49 TD D.Node ΔT= 780s Alt. = 29° Gam. = 0.8691 Dur. = 06m07s	

Annular **2396 Oct 02** Saros 150 21:48 TD D.Node ΔT= 783s Alt. = 81° Gam. = 0.1494 Dur. = 07m12s	**Total** **2397 Mar 29** Saros 155 18:50 TD A.Node ΔT= 785s Alt. = 51° Gam. = 0.6221 Dur. = 03m11s
Total **2399 Aug 02** Saros 142 18:55 TD D.Node ΔT= 793s Alt. = 57° Gam. = 0.5482 Dur. = 05m14s	

Annular **2401 Jan 14** Saros 157 22:15 TD A.Node ΔT= 798s Alt. = 56° Gam. = 0.5616 Dur. = 03m00s	**Total** **2406 Mar 20** Saros 136 17:57 TD D.Node ΔT= 816s Alt. = 43° Gam. = 0.7327 Dur. = 02m03s
Annular **2410 Jan 05** Saros 138 19:32 TD D.Node ΔT= 829s Alt. = 29° Gam. = 0.8749 Dur. = 01m31s	

Plate 030

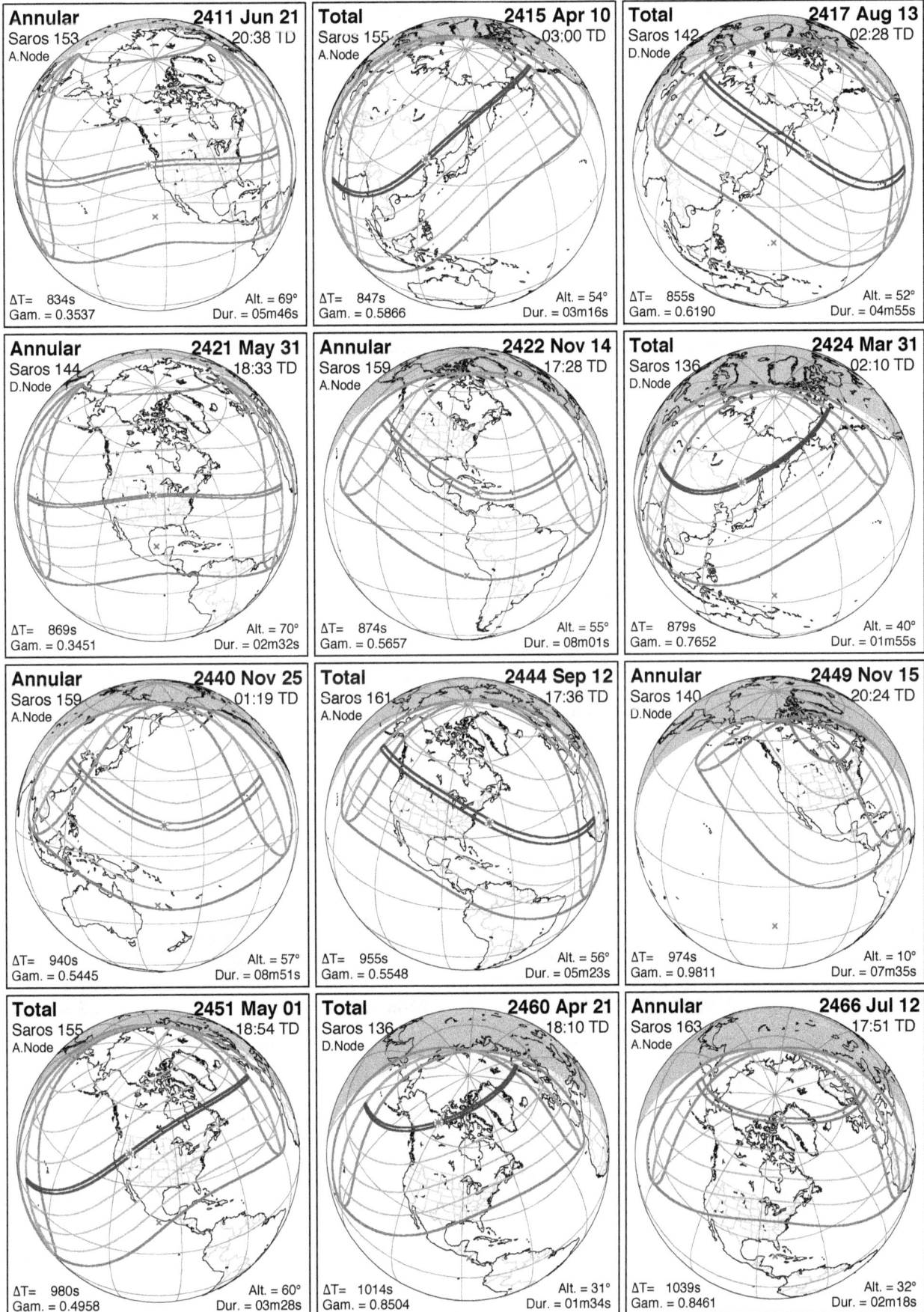

Annular	2411 Jun 21
Saros 153	20:38 TD
A.Node	
ΔT= 834s	Alt. = 69°
Gam. = 0.3537	Dur. = 05m46s

Total	2415 Apr 10
Saros 155	03:00 TD
A.Node	
ΔT= 847s	Alt. = 54°
Gam. = 0.5866	Dur. = 03m16s

Total	2417 Aug 13
Saros 142	02:28 TD
D.Node	
ΔT= 855s	Alt. = 52°
Gam. = 0.6190	Dur. = 04m55s

Annular	2421 May 31
Saros 144	18:33 TD
D.Node	
ΔT= 869s	Alt. = 70°
Gam. = 0.3451	Dur. = 02m32s

Annular	2422 Nov 14
Saros 159	17:28 TD
A.Node	
ΔT= 874s	Alt. = 55°
Gam. = 0.5657	Dur. = 08m01s

Total	2424 Mar 31
Saros 136	02:10 TD
D.Node	
ΔT= 879s	Alt. = 40°
Gam. = 0.7652	Dur. = 01m55s

Annular	2440 Nov 25
Saros 159	01:19 TD
A.Node	
ΔT= 940s	Alt. = 57°
Gam. = 0.5445	Dur. = 08m51s

Total	2444 Sep 12
Saros 161	17:36 TD
A.Node	
ΔT= 955s	Alt. = 56°
Gam. = 0.5548	Dur. = 05m23s

Annular	2449 Nov 15
Saros 140	20:24 TD
D.Node	
ΔT= 974s	Alt. = 10°
Gam. = 0.9811	Dur. = 07m35s

Total	2451 May 01
Saros 155	18:54 TD
A.Node	
ΔT= 980s	Alt. = 60°
Gam. = 0.4958	Dur. = 03m28s

Total	2460 Apr 21
Saros 136	18:10 TD
D.Node	
ΔT= 1014s	Alt. = 31°
Gam. = 0.8504	Dur. = 01m34s

Annular	2466 Jul 12
Saros 163	17:51 TD
A.Node	
ΔT= 1039s	Alt. = 32°
Gam. = 0.8461	Dur. = 02m18s

Plate 031

69

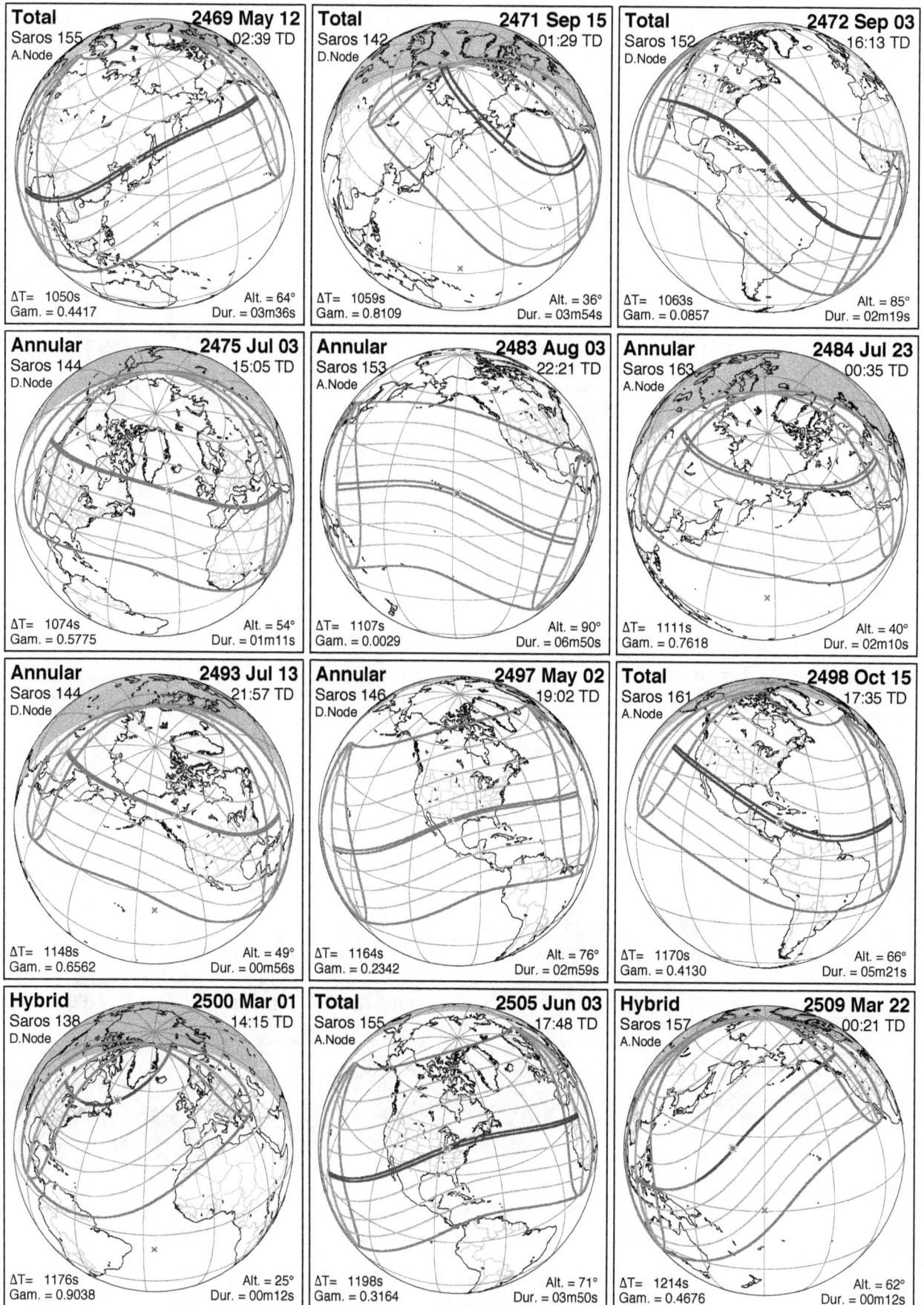

Total 2469 May 12	
Saros 155 · 02:39 TD	
A.Node	
ΔT= 1050s Gam. = 0.4417	Alt. = 64° Dur. = 03m36s

Total 2471 Sep 15	
Saros 142 · 01:29 TD	
D.Node	
ΔT= 1059s Gam. = 0.8109	Alt. = 36° Dur. = 03m54s

Total 2472 Sep 03	
Saros 152 · 16:13 TD	
D.Node	
ΔT= 1063s Gam. = 0.0857	Alt. = 85° Dur. = 02m19s

Annular 2475 Jul 03	
Saros 144 · 15:05 TD	
D.Node	
ΔT= 1074s Gam. = 0.5775	Alt. = 54° Dur. = 01m11s

Annular 2483 Aug 03	
Saros 153 · 22:21 TD	
A.Node	
ΔT= 1107s Gam. = 0.0029	Alt. = 90° Dur. = 06m50s

Annular 2484 Jul 23	
Saros 163 · 00:35 TD	
A.Node	
ΔT= 1111s Gam. = 0.7618	Alt. = 40° Dur. = 02m10s

Annular 2493 Jul 13	
Saros 144 · 21:57 TD	
D.Node	
ΔT= 1148s Gam. = 0.6562	Alt. = 49° Dur. = 00m56s

Annular 2497 May 02	
Saros 146 · 19:02 TD	
D.Node	
ΔT= 1164s Gam. = 0.2342	Alt. = 76° Dur. = 02m59s

Total 2498 Oct 15	
Saros 161 · 17:35 TD	
A.Node	
ΔT= 1170s Gam. = 0.4130	Alt. = 66° Dur. = 05m21s

Hybrid 2500 Mar 01	
Saros 138 · 14:15 TD	
D.Node	
ΔT= 1176s Gam. = 0.9038	Alt. = 25° Dur. = 00m12s

Total 2505 Jun 03	
Saros 155 · 17:48 TD	
A.Node	
ΔT= 1198s Gam. = 0.3164	Alt. = 71° Dur. = 03m50s

Hybrid 2509 Mar 22	
Saros 157 · 00:21 TD	
A.Node	
ΔT= 1214s Gam. = 0.4676	Alt. = 62° Dur. = 00m12s

Plate 032

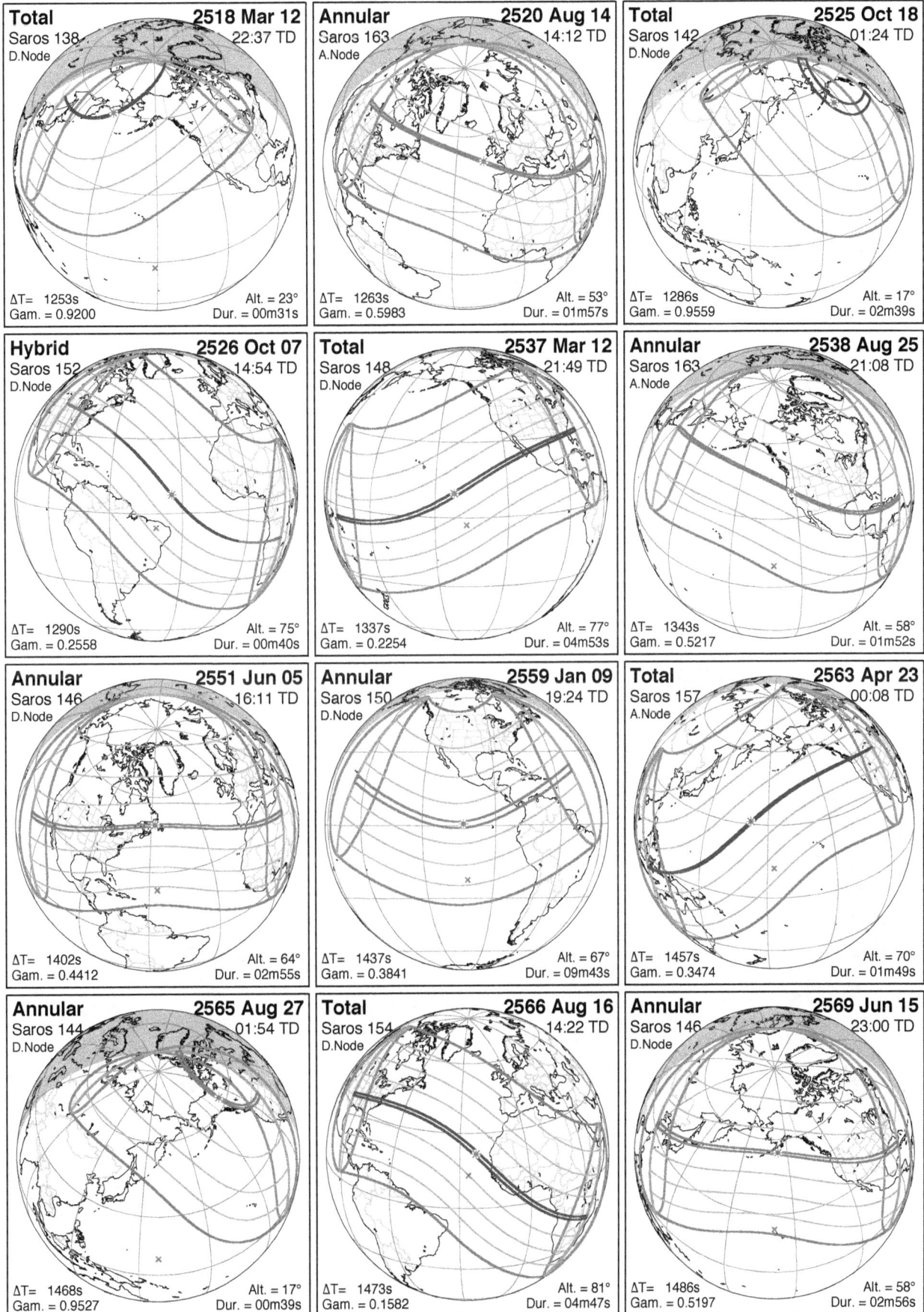

Total	2518 Mar 12
Saros 138	22:37 TD
D.Node	
ΔT= 1253s	Alt. = 23°
Gam. = 0.9200	Dur. = 00m31s

Annular	2520 Aug 14
Saros 163	14:12 TD
A.Node	
ΔT= 1263s	Alt. = 53°
Gam. = 0.5983	Dur. = 01m57s

Total	2525 Oct 18
Saros 142	01:24 TD
D.Node	
ΔT= 1286s	Alt. = 17°
Gam. = 0.9559	Dur. = 02m39s

Hybrid	2526 Oct 07
Saros 152	14:54 TD
D.Node	
ΔT= 1290s	Alt. = 75°
Gam. = 0.2558	Dur. = 00m40s

Total	2537 Mar 12
Saros 148	21:49 TD
D.Node	
ΔT= 1337s	Alt. = 77°
Gam. = 0.2254	Dur. = 04m53s

Annular	2538 Aug 25
Saros 163	21:08 TD
A.Node	
ΔT= 1343s	Alt. = 58°
Gam. = 0.5217	Dur. = 01m52s

Annular	2551 Jun 05
Saros 146	16:11 TD
D.Node	
ΔT= 1402s	Alt. = 64°
Gam. = 0.4412	Dur. = 02m55s

Annular	2559 Jan 09
Saros 150	19:24 TD
D.Node	
ΔT= 1437s	Alt. = 67°
Gam. = 0.3841	Dur. = 09m43s

Total	2563 Apr 23
Saros 157	00:08 TD
A.Node	
ΔT= 1457s	Alt. = 70°
Gam. = 0.3474	Dur. = 01m49s

Annular	2565 Aug 27
Saros 144	01:54 TD
D.Node	
ΔT= 1468s	Alt. = 17°
Gam. = 0.9527	Dur. = 00m39s

Total	2566 Aug 16
Saros 154	14:22 TD
D.Node	
ΔT= 1473s	Alt. = 81°
Gam. = 0.1582	Dur. = 04m47s

Annular	2569 Jun 15
Saros 146	23:00 TD
D.Node	
ΔT= 1486s	Alt. = 58°
Gam. = 0.5197	Dur. = 02m56s

Plate 033

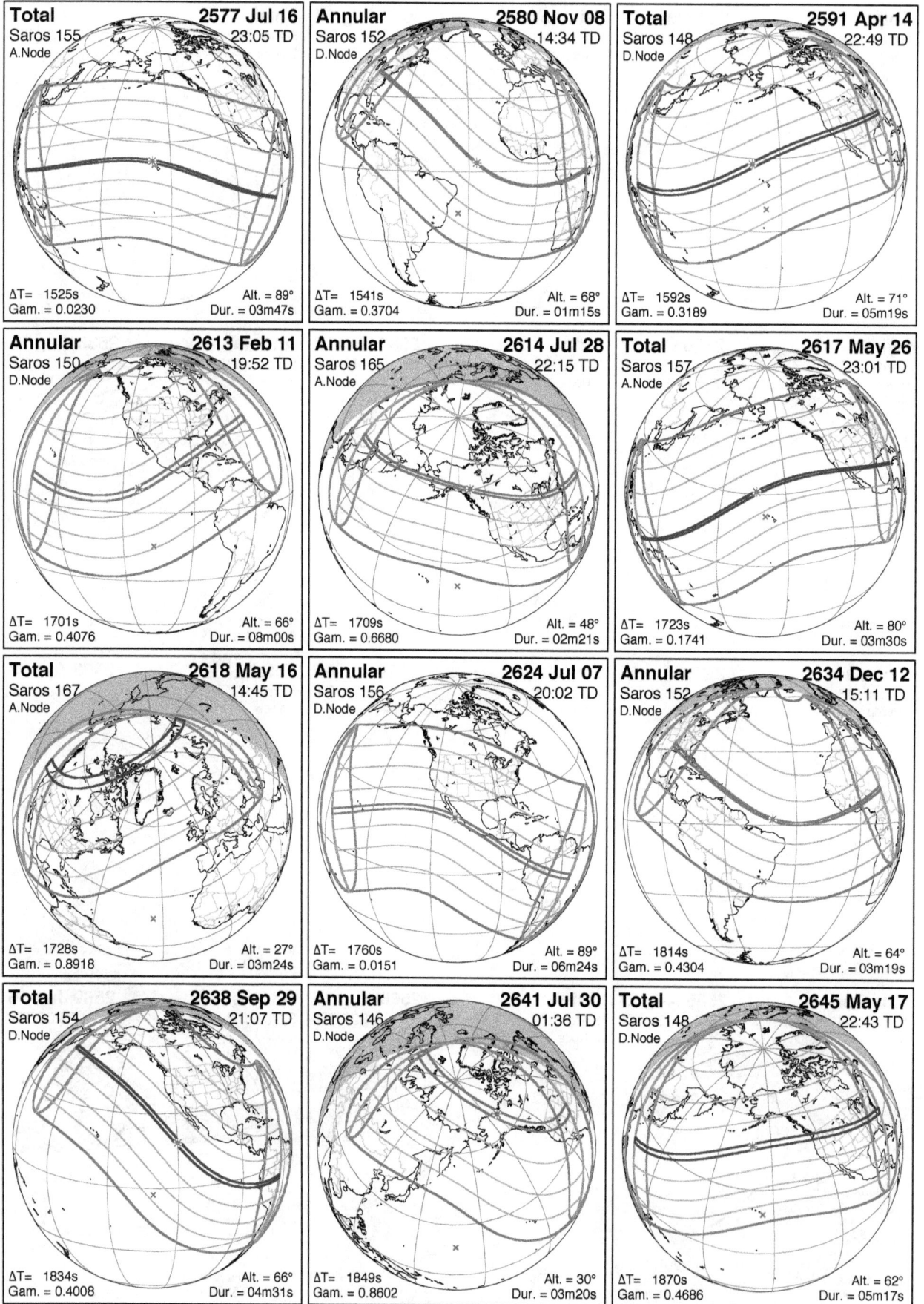

Total	**2577 Jul 16**
Saros 155	23:05 TD
A.Node	

ΔT= 1525s — Alt. = 89°
Gam. = 0.0230 — Dur. = 03m47s

Annular	**2580 Nov 08**
Saros 152	14:34 TD
D.Node	

ΔT= 1541s — Alt. = 68°
Gam. = 0.3704 — Dur. = 01m15s

Total	**2591 Apr 14**
Saros 148	22:49 TD
D.Node	

ΔT= 1592s — Alt. = 71°
Gam. = 0.3189 — Dur. = 05m19s

Annular	**2613 Feb 11**
Saros 150	19:52 TD
D.Node	

ΔT= 1701s — Alt. = 66°
Gam. = 0.4076 — Dur. = 08m00s

Annular	**2614 Jul 28**
Saros 165	22:15 TD
A.Node	

ΔT= 1709s — Alt. = 48°
Gam. = 0.6680 — Dur. = 02m21s

Total	**2617 May 26**
Saros 157	23:01 TD
A.Node	

ΔT= 1723s — Alt. = 80°
Gam. = 0.1741 — Dur. = 03m30s

Total	**2618 May 16**
Saros 167	14:45 TD
A.Node	

ΔT= 1728s — Alt. = 27°
Gam. = 0.8918 — Dur. = 03m24s

Annular	**2624 Jul 07**
Saros 156	20:02 TD
D.Node	

ΔT= 1760s — Alt. = 89°
Gam. = 0.0151 — Dur. = 06m24s

Annular	**2634 Dec 12**
Saros 152	15:11 TD
D.Node	

ΔT= 1814s — Alt. = 64°
Gam. = 0.4304 — Dur. = 03m19s

Total	**2638 Sep 29**
Saros 154	21:07 TD
D.Node	

ΔT= 1834s — Alt. = 66°
Gam. = 0.4008 — Dur. = 04m31s

Annular	**2641 Jul 30**
Saros 146	01:36 TD
D.Node	

ΔT= 1849s — Alt. = 30°
Gam. = 0.8602 — Dur. = 03m20s

Total	**2645 May 17**
Saros 148	22:43 TD
D.Node	

ΔT= 1870s — Alt. = 62°
Gam. = 0.4686 — Dur. = 05m17s

Plate 034

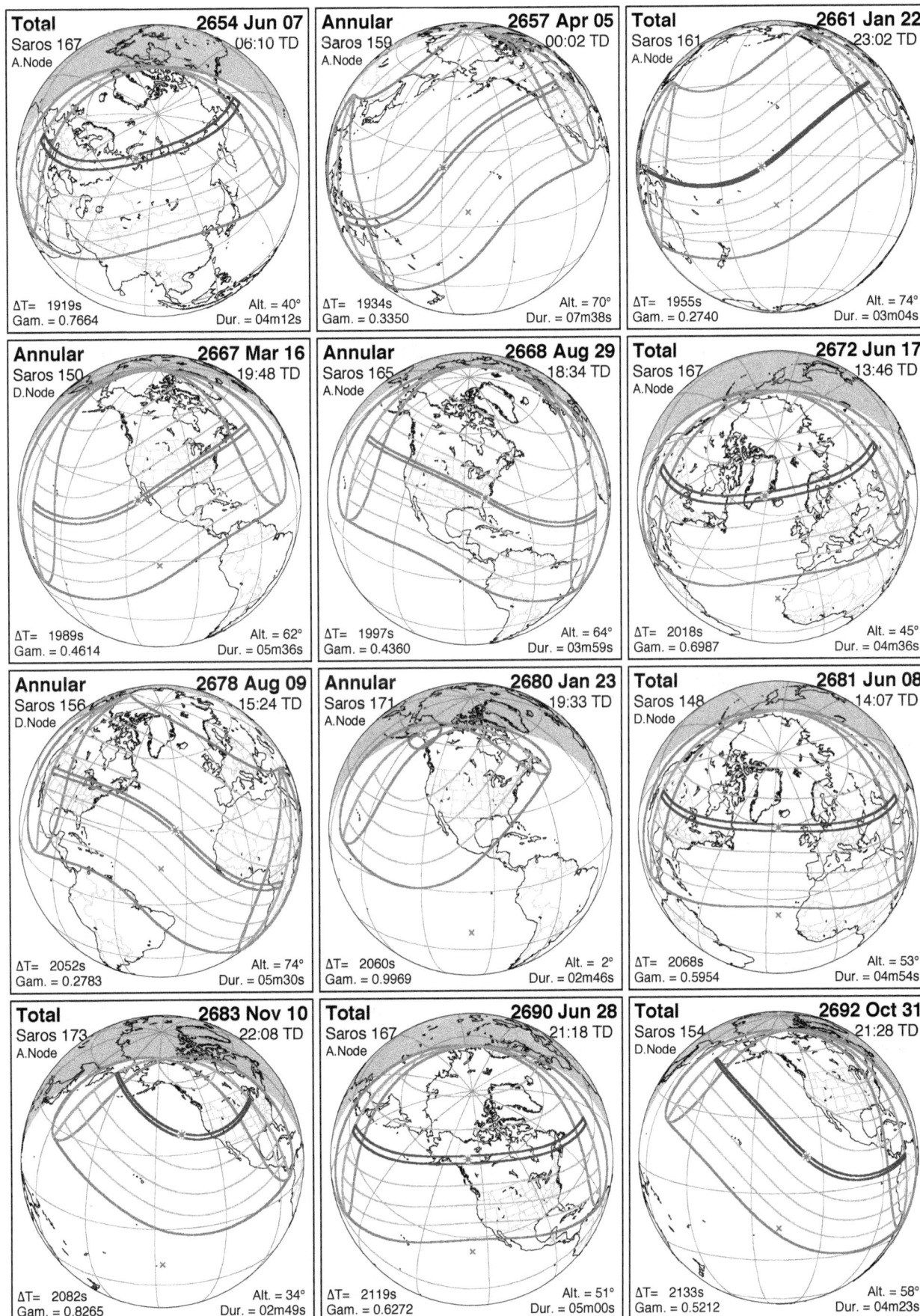

Total	2654 Jun 07
Saros 167	06:10 TD
A.Node	
ΔT= 1919s	Alt. = 40°
Gam. = 0.7664	Dur. = 04m12s

Annular	2657 Apr 05
Saros 159	00:02 TD
A.Node	
ΔT= 1934s	Alt. = 70°
Gam. = 0.3350	Dur. = 07m38s

Total	2661 Jan 22
Saros 161	23:02 TD
A.Node	
ΔT= 1955s	Alt. = 74°
Gam. = 0.2740	Dur. = 03m04s

Annular	2667 Mar 16
Saros 150	19:48 TD
D.Node	
ΔT= 1989s	Alt. = 62°
Gam. = 0.4614	Dur. = 05m36s

Annular	2668 Aug 29
Saros 165	18:34 TD
A.Node	
ΔT= 1997s	Alt. = 64°
Gam. = 0.4360	Dur. = 03m59s

Total	2672 Jun 17
Saros 167	13:46 TD
A.Node	
ΔT= 2018s	Alt. = 45°
Gam. = 0.6987	Dur. = 04m36s

Annular	2678 Aug 09
Saros 156	15:24 TD
D.Node	
ΔT= 2052s	Alt. = 74°
Gam. = 0.2783	Dur. = 05m30s

Annular	2680 Jan 23
Saros 171	19:33 TD
A.Node	
ΔT= 2060s	Alt. = 2°
Gam. = 0.9969	Dur. = 02m46s

Total	2681 Jun 08
Saros 148	14:07 TD
D.Node	
ΔT= 2068s	Alt. = 53°
Gam. = 0.5954	Dur. = 04m54s

Total	2683 Nov 10
Saros 173	22:08 TD
A.Node	
ΔT= 2082s	Alt. = 34°
Gam. = 0.8265	Dur. = 02m49s

Total	2690 Jun 28
Saros 167	21:18 TD
A.Node	
ΔT= 2119s	Alt. = 51°
Gam. = 0.6272	Dur. = 05m00s

Total	2692 Oct 31
Saros 154	21:28 TD
D.Node	
ΔT= 2133s	Alt. = 58°
Gam. = 0.5212	Dur. = 04m23s

Plate 035

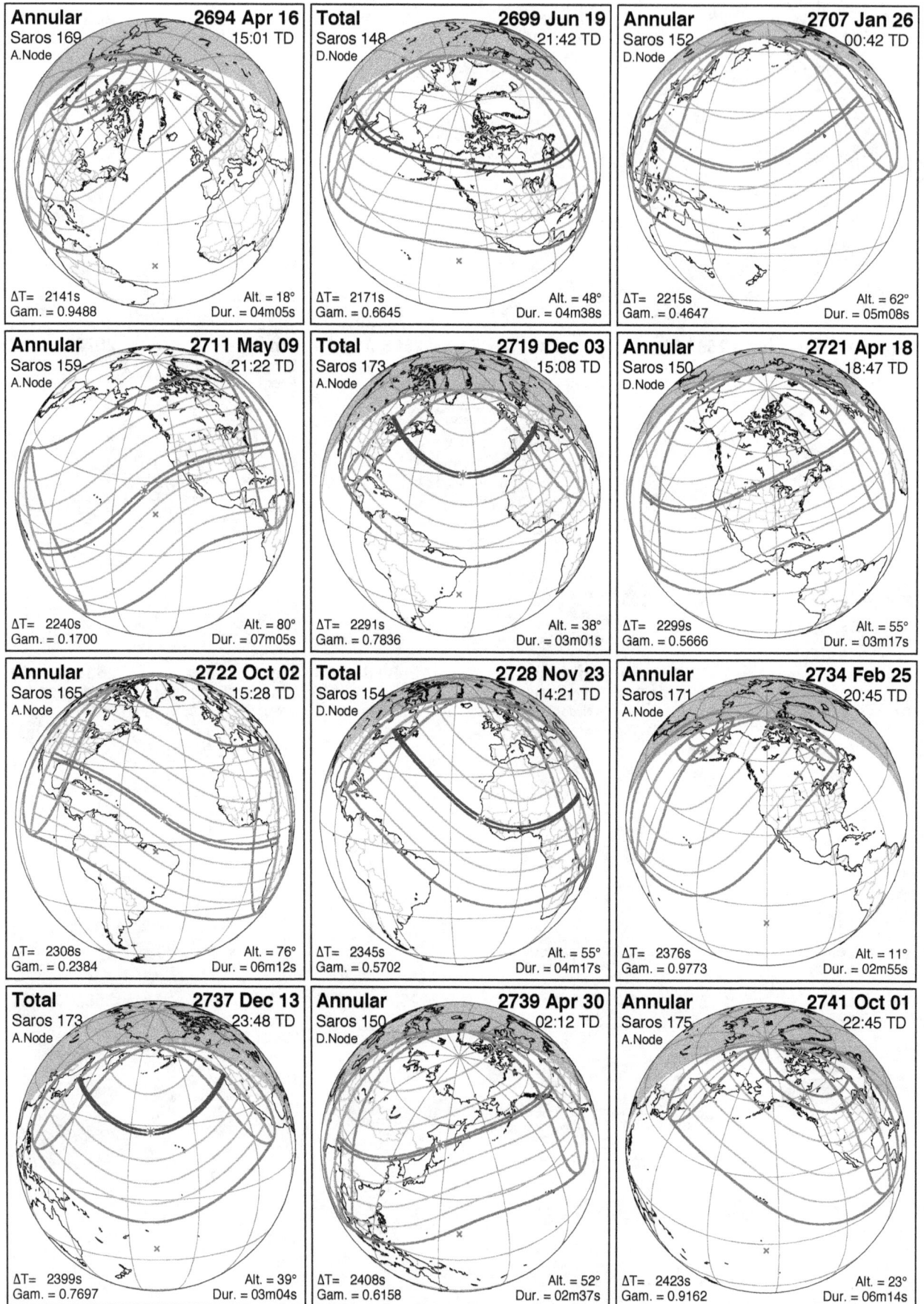

Annular **2694 Apr 16** Saros 169 15:01 TD A.Node ΔT= 2141s Alt. = 18° Gam. = 0.9488 Dur. = 04m05s	**Total** **2699 Jun 19** Saros 148 21:42 TD D.Node ΔT= 2171s Alt. = 48° Gam. = 0.6645 Dur. = 04m38s

Annular   **2707 Jan 26**
 Saros 152   00:42 TD
 D.Node

 ΔT= 2215s   Alt. = 62°
 Gam. = 0.4647   Dur. = 05m08s

Annular   **2711 May 09**
 Saros 159   21:22 TD
 A.Node

 ΔT= 2240s   Alt. = 80°
 Gam. = 0.1700   Dur. = 07m05s

Total   **2719 Dec 03**
 Saros 173   15:08 TD
 A.Node

 ΔT= 2291s   Alt. = 38°
 Gam. = 0.7836   Dur. = 03m01s

Annular   **2721 Apr 18**
 Saros 150   18:47 TD
 D.Node

 ΔT= 2299s   Alt. = 55°
 Gam. = 0.5666   Dur. = 03m17s

Annular   **2722 Oct 02**
 Saros 165   15:28 TD
 A.Node

 ΔT= 2308s   Alt. = 76°
 Gam. = 0.2384   Dur. = 06m12s

Total   **2728 Nov 23**
 Saros 154   14:21 TD
 D.Node

 ΔT= 2345s   Alt. = 55°
 Gam. = 0.5702   Dur. = 04m17s

Annular   **2734 Feb 25**
 Saros 171   20:45 TD
 A.Node

 ΔT= 2376s   Alt. = 11°
 Gam. = 0.9773   Dur. = 02m55s

Total   **2737 Dec 13**
 Saros 173   23:48 TD
 A.Node

 ΔT= 2399s   Alt. = 39°
 Gam. = 0.7697   Dur. = 03m04s

Annular   **2739 Apr 30**
 Saros 150   02:12 TD
 D.Node

 ΔT= 2408s   Alt. = 52°
 Gam. = 0.6158   Dur. = 02m37s

Annular   **2741 Oct 01**
 Saros 175   22:45 TD
 A.Node

 ΔT= 2423s   Alt. = 23°
 Gam. = 0.9162   Dur. = 06m14s

Plate 036

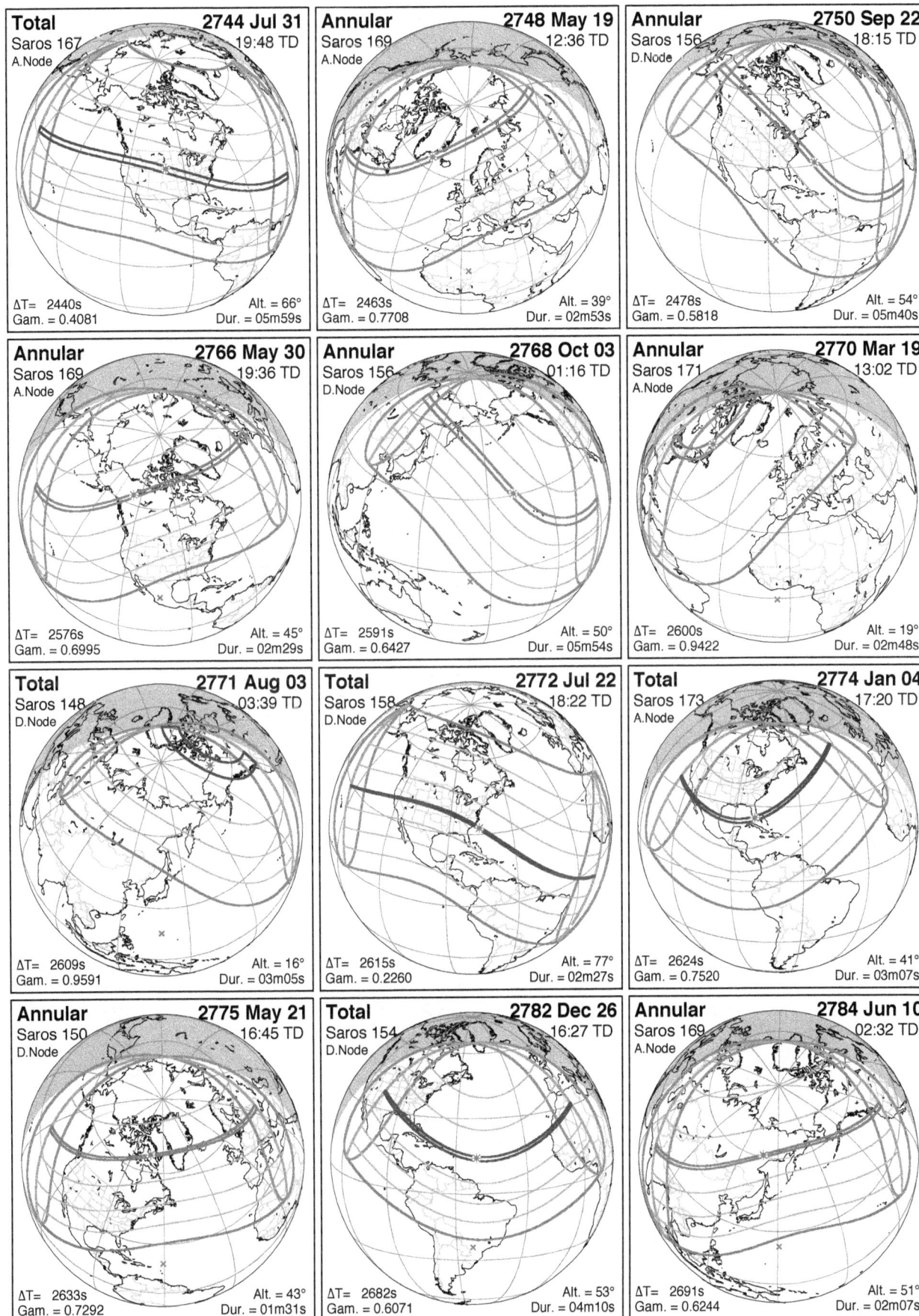

Total **2744 Jul 31**	
Saros 167 19:48 TD	
A.Node	
ΔT= 2440s Alt. = 66°	
Gam. = 0.4081 Dur. = 05m59s	

Annular **2748 May 19**	
Saros 169 12:36 TD	
A.Node	
ΔT= 2463s Alt. = 39°	
Gam. = 0.7708 Dur. = 02m53s	

Annular **2750 Sep 22**	
Saros 156 18:15 TD	
D.Node	
ΔT= 2478s Alt. = 54°	
Gam. = 0.5818 Dur. = 05m40s	

Annular **2766 May 30**	
Saros 169 19:36 TD	
A.Node	
ΔT= 2576s Alt. = 45°	
Gam. = 0.6995 Dur. = 02m29s	

Annular **2768 Oct 03**	
Saros 156 01:16 TD	
D.Node	
ΔT= 2591s Alt. = 50°	
Gam. = 0.6427 Dur. = 05m54s	

Annular **2770 Mar 19**	
Saros 171 13:02 TD	
A.Node	
ΔT= 2600s Alt. = 19°	
Gam. = 0.9422 Dur. = 02m48s	

Total **2771 Aug 03**	
Saros 148 03:39 TD	
D.Node	
ΔT= 2609s Alt. = 16°	
Gam. = 0.9591 Dur. = 03m05s	

Total **2772 Jul 22**	
Saros 158 18:22 TD	
D.Node	
ΔT= 2615s Alt. = 77°	
Gam. = 0.2260 Dur. = 02m27s	

Total **2774 Jan 04**	
Saros 173 17:20 TD	
A.Node	
ΔT= 2624s Alt. = 41°	
Gam. = 0.7520 Dur. = 03m07s	

Annular **2775 May 21**	
Saros 150 16:45 TD	
D.Node	
ΔT= 2633s Alt. = 43°	
Gam. = 0.7292 Dur. = 01m31s	

Total **2782 Dec 26**	
Saros 154 16:27 TD	
D.Node	
ΔT= 2682s Alt. = 53°	
Gam. = 0.6071 Dur. = 04m10s	

Annular **2784 Jun 10**	
Saros 169 02:32 TD	
A.Node	
ΔT= 2691s Alt. = 51°	
Gam. = 0.6244 Dur. = 02m07s	

Plate 037

75

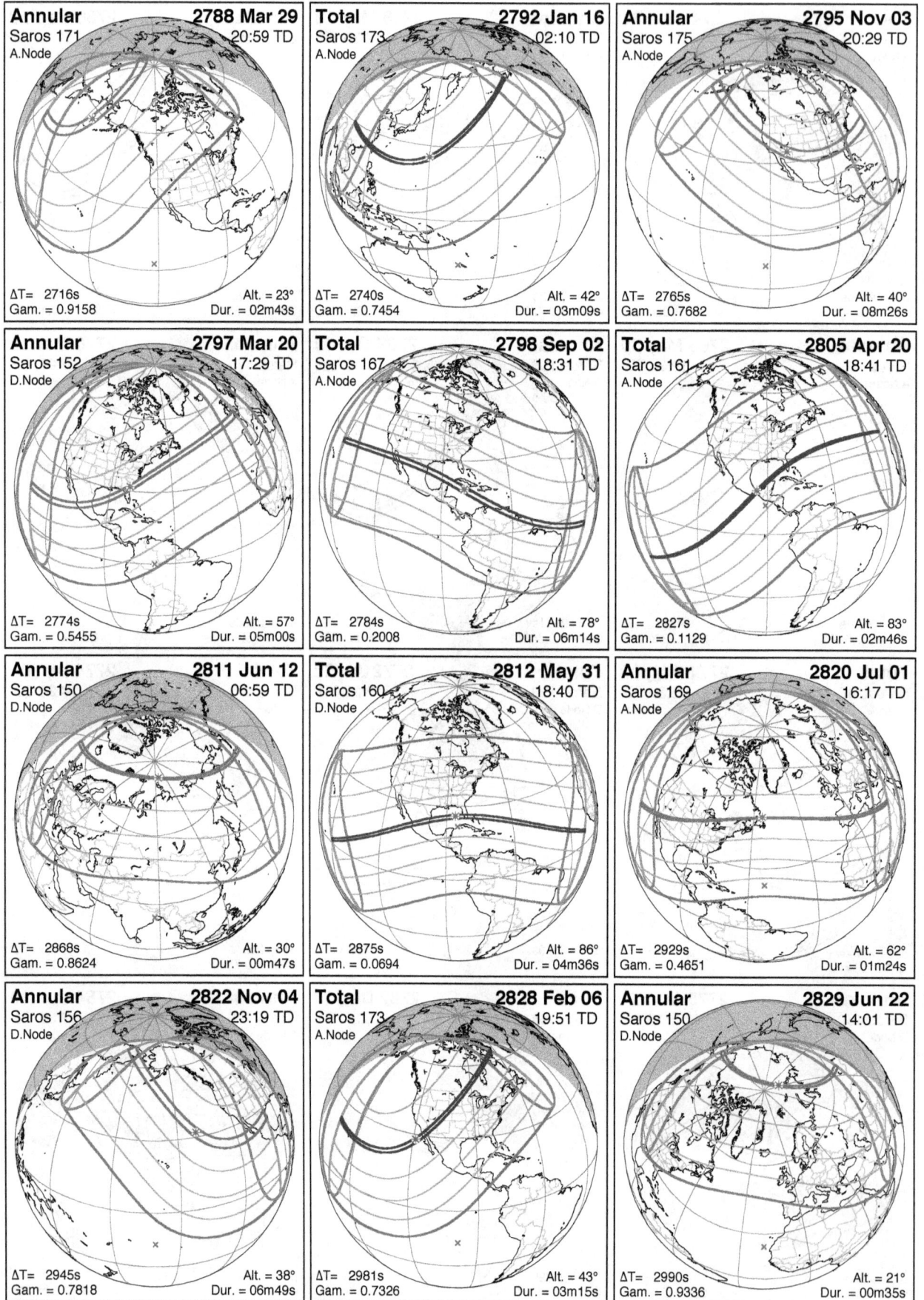

Annular 2788 Mar 29 Saros 171 20:59 TD A.Node ΔT= 2716s Alt. = 23° Gam. = 0.9158 Dur. = 02m43s	**Total** 2792 Jan 16 Saros 173 02:10 TD A.Node ΔT= 2740s Alt. = 42° Gam. = 0.7454 Dur. = 03m09s	**Annular** 2795 Nov 03 Saros 175 20:29 TD A.Node ΔT= 2765s Alt. = 40° Gam. = 0.7682 Dur. = 08m26s
Annular 2797 Mar 20 Saros 152 17:29 TD D.Node ΔT= 2774s Alt. = 57° Gam. = 0.5455 Dur. = 05m00s	**Total** 2798 Sep 02 Saros 167 18:31 TD A.Node ΔT= 2784s Alt. = 78° Gam. = 0.2008 Dur. = 06m14s	**Total** 2805 Apr 20 Saros 161 18:41 TD A.Node ΔT= 2827s Alt. = 83° Gam. = 0.1129 Dur. = 02m46s
Annular 2811 Jun 12 Saros 150 06:59 TD D.Node ΔT= 2868s Alt. = 30° Gam. = 0.8624 Dur. = 00m47s	**Total** 2812 May 31 Saros 160 18:40 TD D.Node ΔT= 2875s Alt. = 86° Gam. = 0.0694 Dur. = 04m36s	**Annular** 2820 Jul 01 Saros 169 16:17 TD A.Node ΔT= 2929s Alt. = 62° Gam. = 0.4651 Dur. = 01m24s
Annular 2822 Nov 04 Saros 156 23:19 TD D.Node ΔT= 2945s Alt. = 38° Gam. = 0.7818 Dur. = 06m49s	**Total** 2828 Feb 06 Saros 173 19:51 TD A.Node ΔT= 2981s Alt. = 43° Gam. = 0.7326 Dur. = 03m15s	**Annular** 2829 Jun 22 Saros 150 14:01 TD D.Node ΔT= 2990s Alt. = 21° Gam. = 0.9336 Dur. = 00m35s

Plate 038

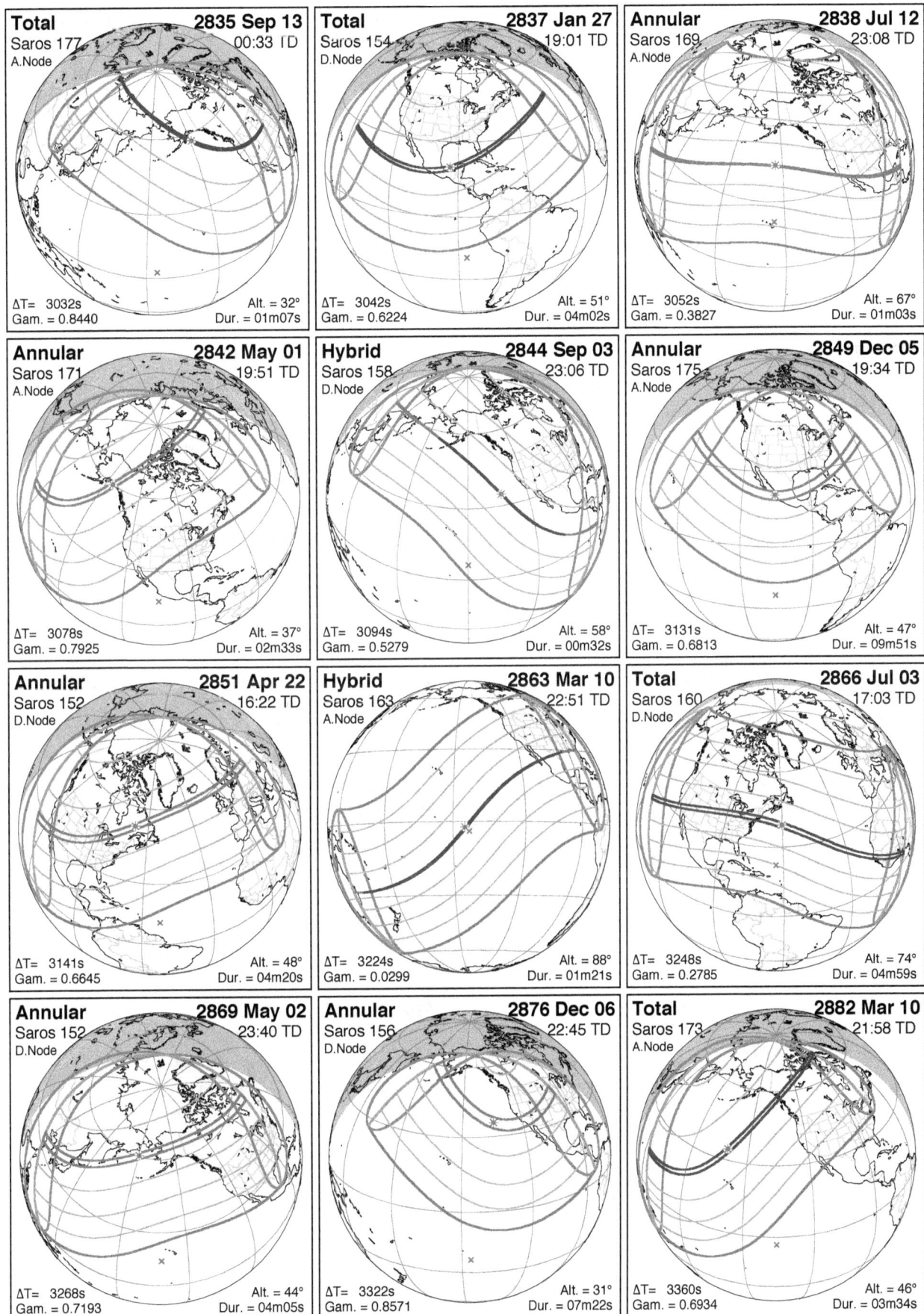

Total	2835 Sep 13
Saros 177	00:33 TD
A.Node	
ΔT= 3032s	Alt. = 32°
Gam. = 0.8440	Dur. = 01m07s

Total	2837 Jan 27
Saros 154	19:01 TD
D.Node	
ΔT= 3042s	Alt. = 51°
Gam. = 0.6224	Dur. = 04m02s

Annular	2838 Jul 12
Saros 169	23:08 TD
A.Node	
ΔT= 3052s	Alt. = 67°
Gam. = 0.3827	Dur. = 01m03s

Annular	2842 May 01
Saros 171	19:51 TD
A.Node	
ΔT= 3078s	Alt. = 37°
Gam. = 0.7925	Dur. = 02m33s

Hybrid	2844 Sep 03
Saros 158	23:06 TD
D.Node	
ΔT= 3094s	Alt. = 58°
Gam. = 0.5279	Dur. = 00m32s

Annular	2849 Dec 05
Saros 175	19:34 TD
A.Node	
ΔT= 3131s	Alt. = 47°
Gam. = 0.6813	Dur. = 09m51s

Annular	2851 Apr 22
Saros 152	16:22 TD
D.Node	
ΔT= 3141s	Alt. = 48°
Gam. = 0.6645	Dur. = 04m20s

Hybrid	2863 Mar 10
Saros 163	22:51 TD
A.Node	
ΔT= 3224s	Alt. = 88°
Gam. = 0.0299	Dur. = 01m21s

Total	2866 Jul 03
Saros 160	17:03 TD
D.Node	
ΔT= 3248s	Alt. = 74°
Gam. = 0.2785	Dur. = 04m59s

Annular	2869 May 02
Saros 152	23:40 TD
D.Node	
ΔT= 3268s	Alt. = 44°
Gam. = 0.7193	Dur. = 04m05s

Annular	2876 Dec 06
Saros 156	22:45 TD
D.Node	
ΔT= 3322s	Alt. = 31°
Gam. = 0.8571	Dur. = 07m22s

Total	2882 Mar 10
Saros 173	21:58 TD
A.Node	
ΔT= 3360s	Alt. = 46°
Gam. = 0.6934	Dur. = 03m34s

Plate 039

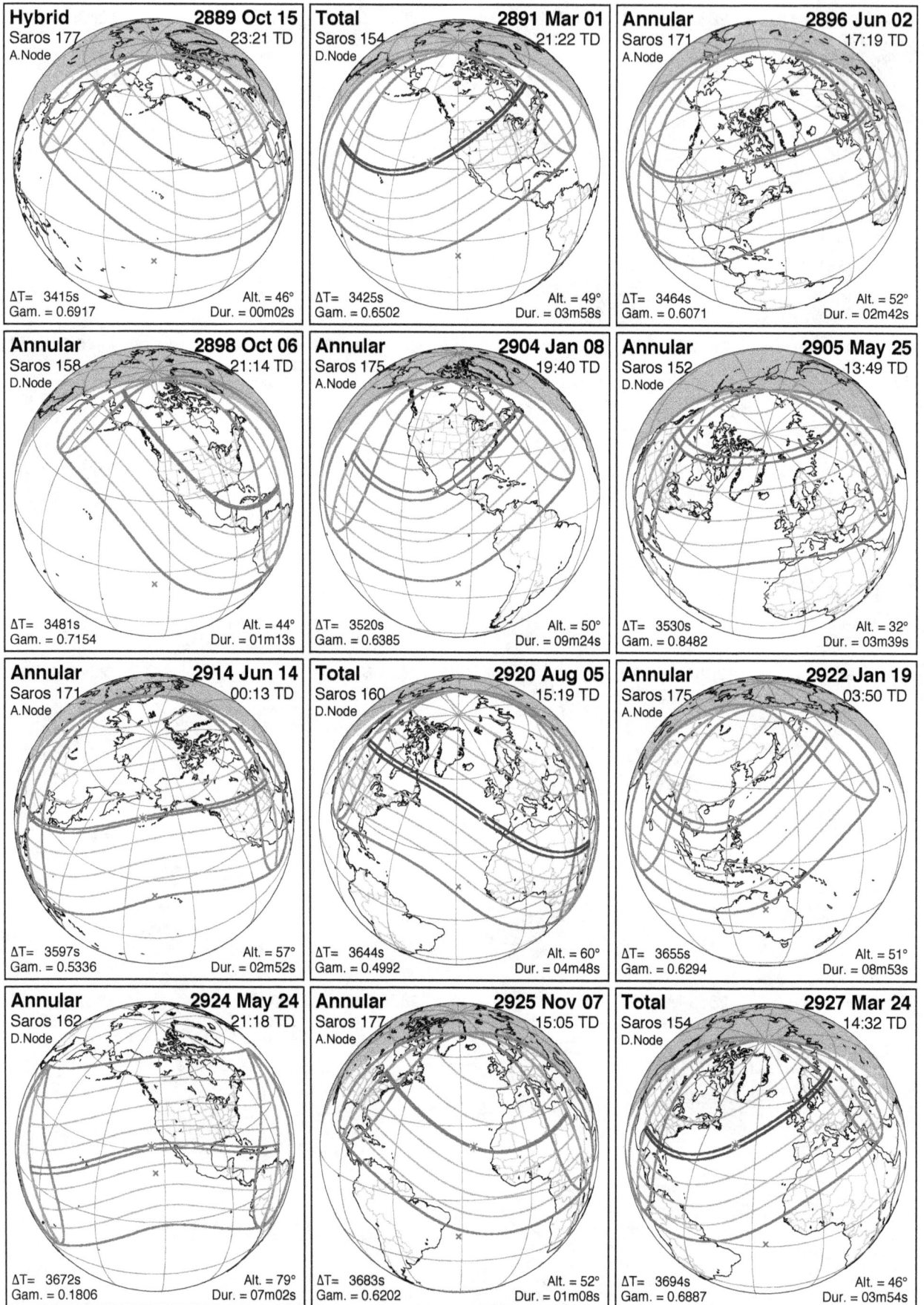

Hybrid **2889 Oct 15**	
Saros 177 23:21 TD	
A.Node	
ΔT= 3415s Alt. = 46°	
Gam. = 0.6917 Dur. = 00m02s	

Hybrid   **2889 Oct 15**
Saros 177   23:21 TD
A.Node
ΔT= 3415s   Alt. = 46°
Gam. = 0.6917   Dur. = 00m02s

Total   **2891 Mar 01**
Saros 154   21:22 TD
D.Node
ΔT= 3425s   Alt. = 49°
Gam. = 0.6502   Dur. = 03m58s

Annular   **2896 Jun 02**
Saros 171   17:19 TD
A.Node
ΔT= 3464s   Alt. = 52°
Gam. = 0.6071   Dur. = 02m42s

Annular   **2898 Oct 06**
Saros 158   21:14 TD
D.Node
ΔT= 3481s   Alt. = 44°
Gam. = 0.7154   Dur. = 01m13s

Annular   **2904 Jan 08**
Saros 175   19:40 TD
A.Node
ΔT= 3520s   Alt. = 50°
Gam. = 0.6385   Dur. = 09m24s

Annular   **2905 May 25**
Saros 152   13:49 TD
D.Node
ΔT= 3530s   Alt. = 32°
Gam. = 0.8482   Dur. = 03m39s

Annular   **2914 Jun 14**
Saros 171   00:13 TD
A.Node
ΔT= 3597s   Alt. = 57°
Gam. = 0.5336   Dur. = 02m52s

Total   **2920 Aug 05**
Saros 160   15:19 TD
D.Node
ΔT= 3644s   Alt. = 60°
Gam. = 0.4992   Dur. = 04m48s

Annular   **2922 Jan 19**
Saros 175   03:50 TD
A.Node
ΔT= 3655s   Alt. = 51°
Gam. = 0.6294   Dur. = 08m53s

Annular   **2924 May 24**
Saros 162   21:18 TD
D.Node
ΔT= 3672s   Alt. = 79°
Gam. = 0.1806   Dur. = 07m02s

Annular   **2925 Nov 07**
Saros 177   15:05 TD
A.Node
ΔT= 3683s   Alt. = 52°
Gam. = 0.6202   Dur. = 01m08s

Total   **2927 Mar 24**
Saros 154   14:32 TD
D.Node
ΔT= 3694s   Alt. = 46°
Gam. = 0.6887   Dur. = 03m54s

Plate 040

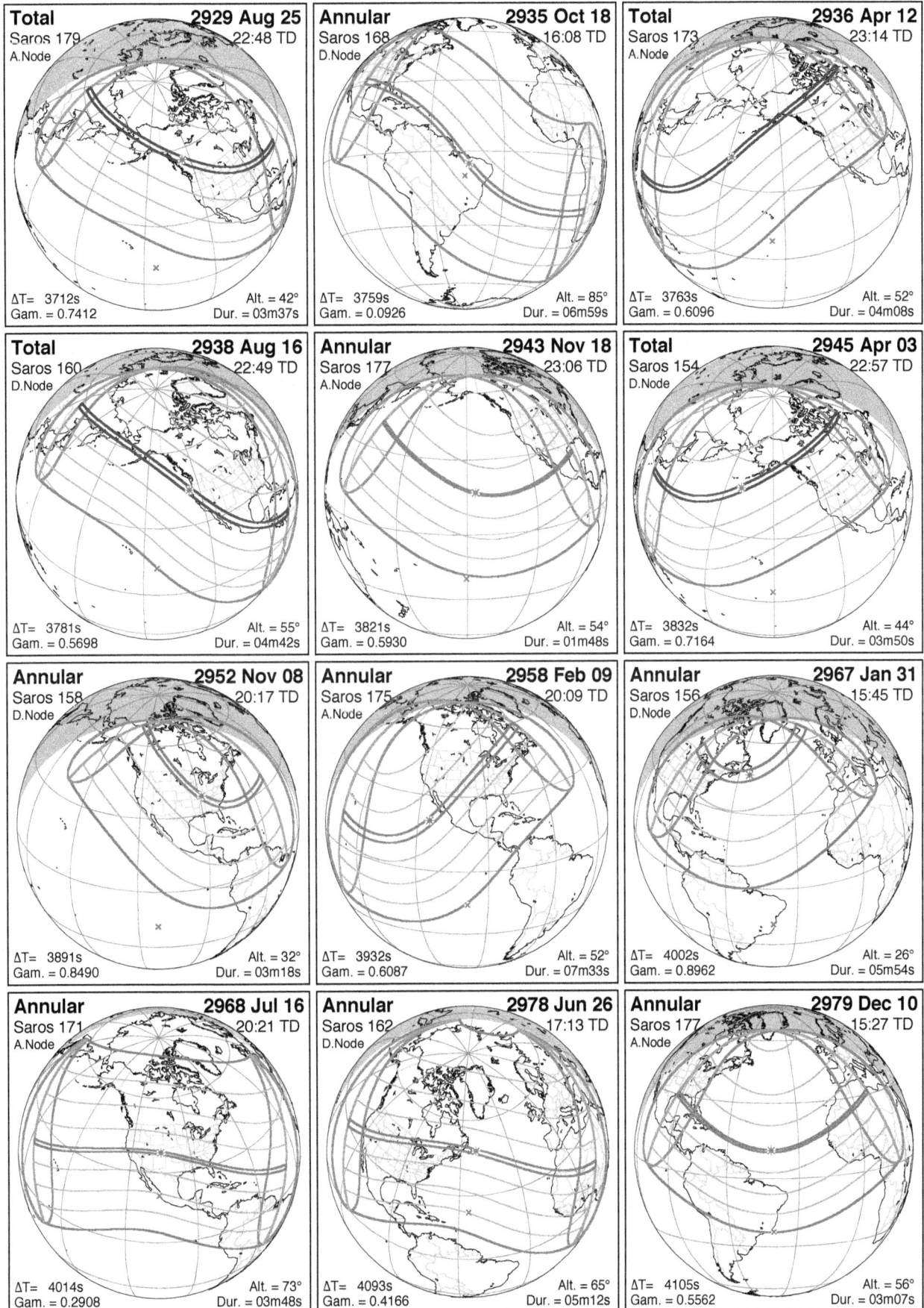

Total	2929 Aug 25
Saros 179	22:48 TD
A.Node	
ΔT= 3712s	Alt. = 42°
Gam. = 0.7412	Dur. = 03m37s

Annular	2935 Oct 18
Saros 168	16:08 TD
D.Node	
ΔT= 3759s	Alt. = 85°
Gam. = 0.0926	Dur. = 06m59s

Total	2936 Apr 12
Saros 173	23:14 TD
A.Node	
ΔT= 3763s	Alt. = 52°
Gam. = 0.6096	Dur. = 04m08s

Total	2938 Aug 16
Saros 160	22:49 TD
D.Node	
ΔT= 3781s	Alt. = 55°
Gam. = 0.5698	Dur. = 04m42s

Annular	2943 Nov 18
Saros 177	23:06 TD
A.Node	
ΔT= 3821s	Alt. = 54°
Gam. = 0.5930	Dur. = 01m48s

Total	2945 Apr 03
Saros 154	22:57 TD
D.Node	
ΔT= 3832s	Alt. = 44°
Gam. = 0.7164	Dur. = 03m50s

Annular	2952 Nov 08
Saros 158	20:17 TD
D.Node	
ΔT= 3891s	Alt. = 32°
Gam. = 0.8490	Dur. = 03m18s

Annular	2958 Feb 09
Saros 175	20:09 TD
A.Node	
ΔT= 3932s	Alt. = 52°
Gam. = 0.6087	Dur. = 07m33s

Annular	2967 Jan 31
Saros 156	15:45 TD
D.Node	
ΔT= 4002s	Alt. = 26°
Gam. = 0.8962	Dur. = 05m54s

Annular	2968 Jul 16
Saros 171	20:21 TD
A.Node	
ΔT= 4014s	Alt. = 73°
Gam. = 0.2908	Dur. = 03m48s

Annular	2978 Jun 26
Saros 162	17:13 TD
D.Node	
ΔT= 4093s	Alt. = 65°
Gam. = 0.4166	Dur. = 05m12s

Annular	2979 Dec 10
Saros 177	15:27 TD
A.Node	
ΔT= 4105s	Alt. = 56°
Gam. = 0.5562	Dur. = 03m07s

Plate 041

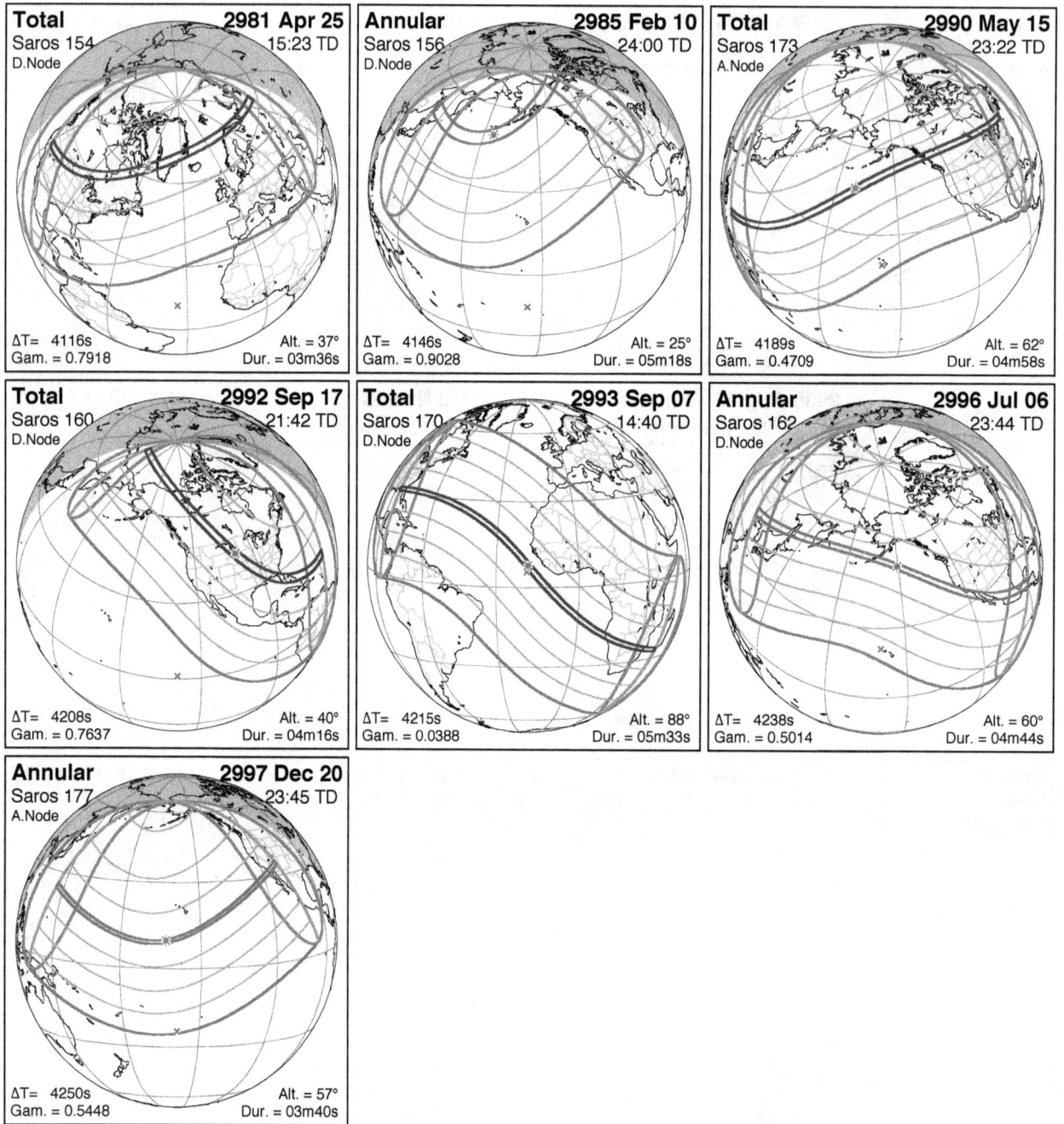

Total 2981 Apr 25	
Saros 154 15:23 TD	
D.Node	
ΔT= 4116s Alt. = 37°	
Gam. = 0.7918 Dur. = 03m36s	

Annular 2985 Feb 10	
Saros 156 24:00 TD	
D.Node	
ΔT= 4146s Alt. = 25°	
Gam. = 0.9028 Dur. = 05m18s	

Total 2990 May 15	
Saros 173 23:22 TD	
A.Node	
ΔT= 4189s Alt. = 62°	
Gam. = 0.4709 Dur. = 04m58s	

Total 2992 Sep 17	
Saros 160 21:42 TD	
D.Node	
ΔT= 4208s Alt. = 40°	
Gam. = 0.7637 Dur. = 04m16s	

Total 2993 Sep 07	
Saros 170 14:40 TD	
D.Node	
ΔT= 4215s Alt. = 88°	
Gam. = 0.0388 Dur. = 05m33s	

Annular 2996 Jul 06	
Saros 162 23:44 TD	
D.Node	
ΔT= 4238s Alt. = 60°	
Gam. = 0.5014 Dur. = 04m44s	

Annular 2997 Dec 20	
Saros 177 23:45 TD	
A.Node	
ΔT= 4250s Alt. = 57°	
Gam. = 0.5448 Dur. = 03m40s	

Plate 042

Appendix C

Maps of Central Solar Eclipses in the Lower 48 States of the USA

Map 01: Central Solar Eclipses of 1001–1050 CE

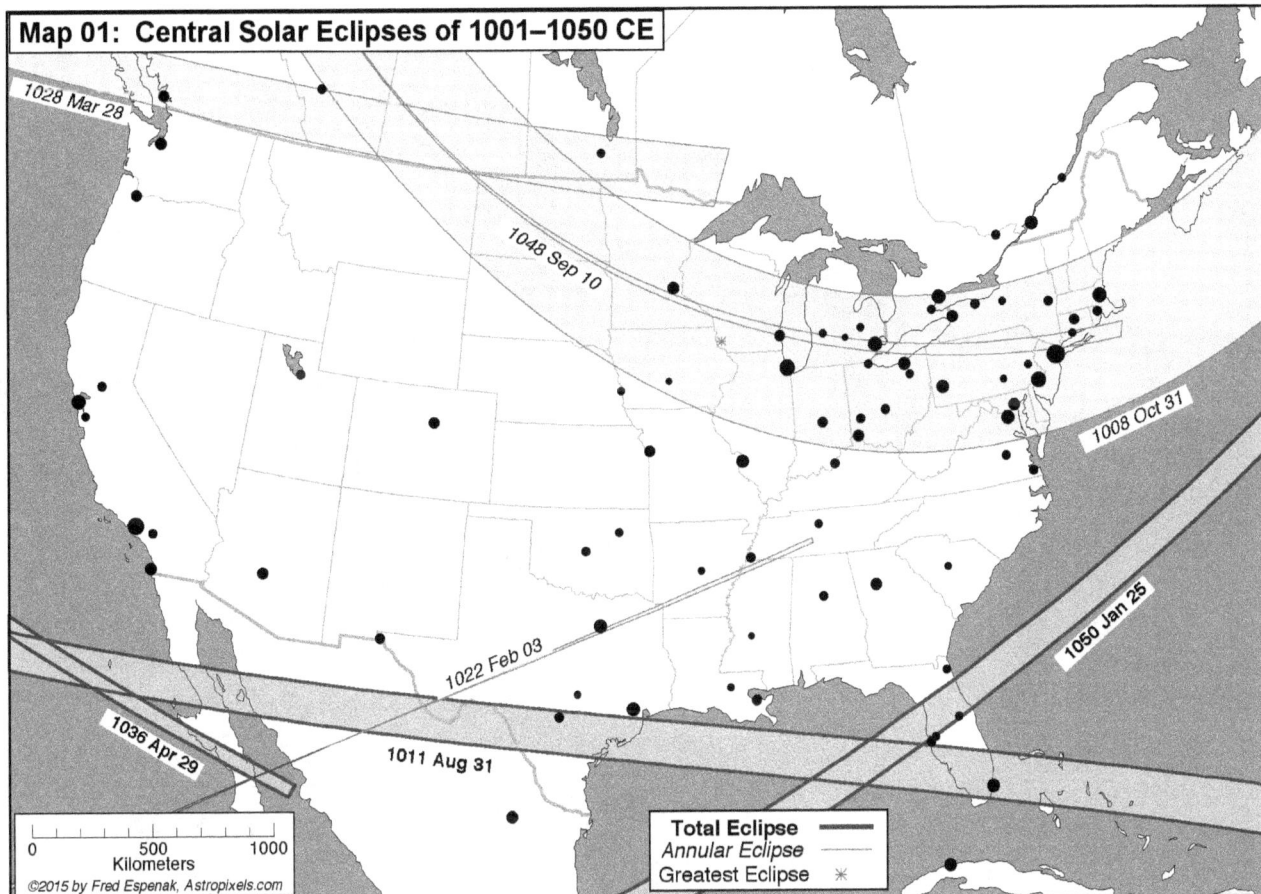

1028 Mar 28
1048 Sep 10
1008 Oct 31
1050 Jan 25
1022 Feb 03
1036 Apr 29
1011 Aug 31

0   500   1000
Kilometers
©2015 by Fred Espenak, Astropixels.com

Total Eclipse
Annular Eclipse
Greatest Eclipse

Map 02: Central Solar Eclipses of 1051–1110 CE

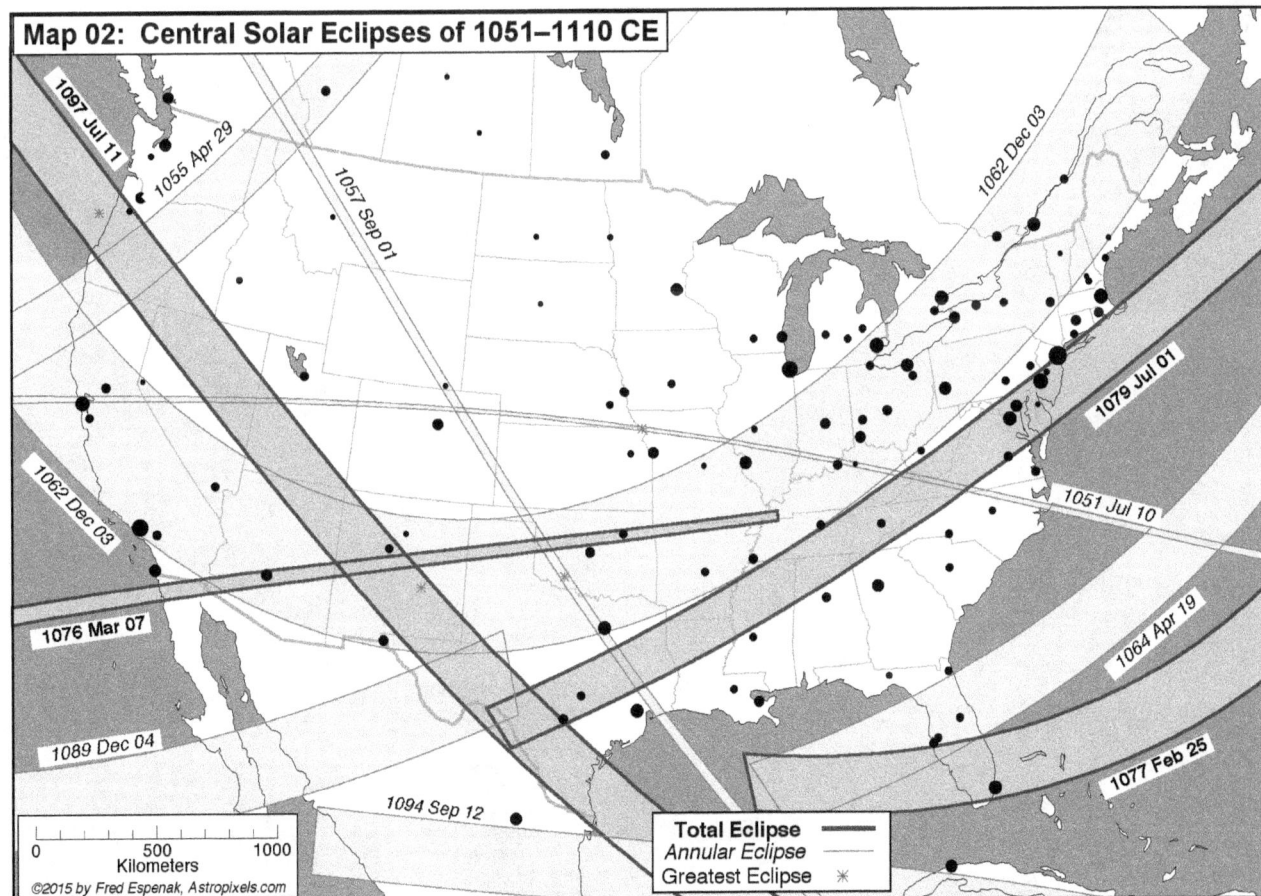

1097 Jul 11
1055 Apr 29
1057 Sep 01
1062 Dec 03
1079 Jul 01
1062 Dec 03
1051 Jul 10
1076 Mar 07
1064 Apr 19
1089 Dec 04
1094 Sep 12
1077 Feb 25

0   500   1000
Kilometers
©2015 by Fred Espenak, Astropixels.com

Total Eclipse
Annular Eclipse
Greatest Eclipse

Map 03: Central Solar Eclipses of 1101–1150 CE

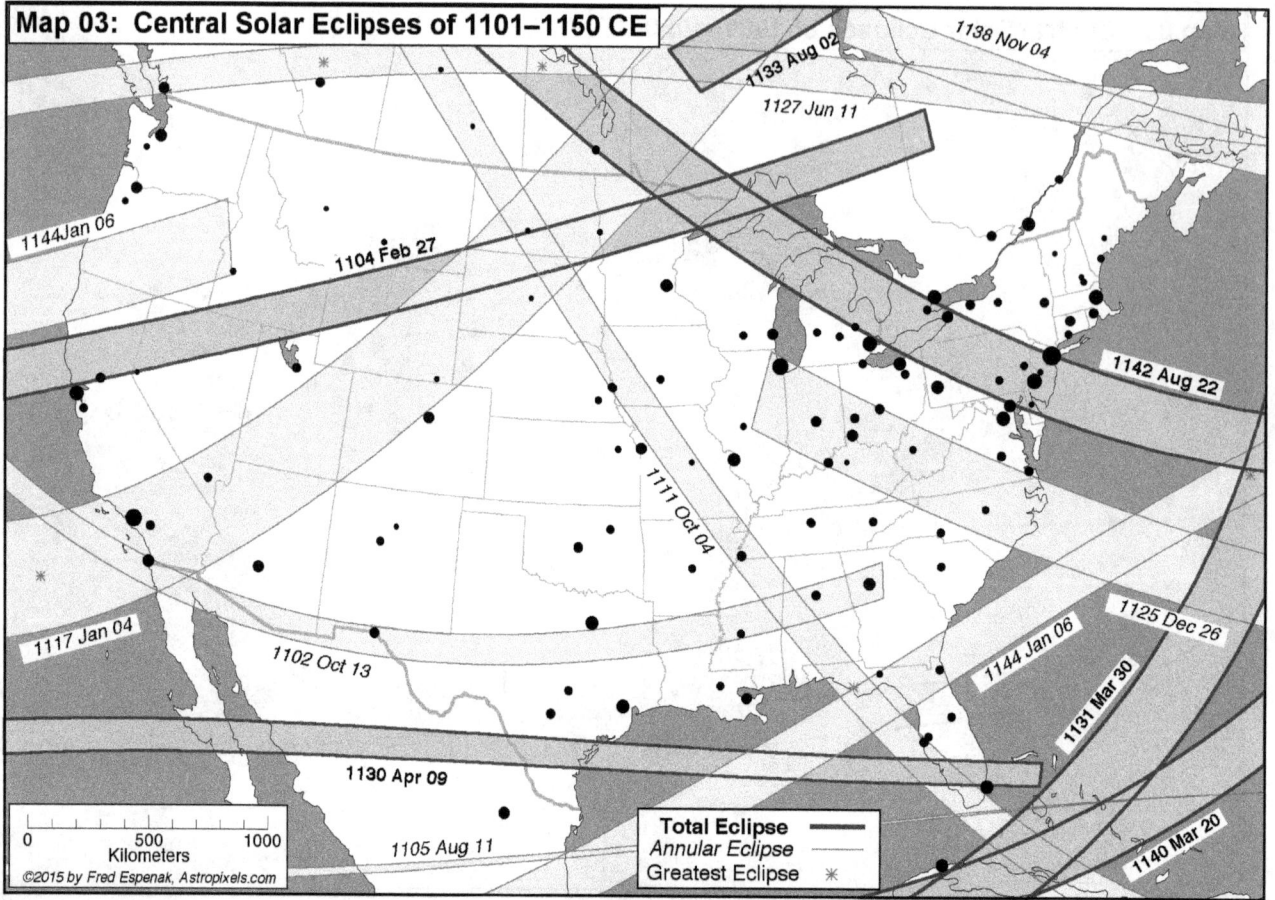

1133 Aug 02
1138 Nov 04
1127 Jun 11
1144 Jan 06
1104 Feb 27
1142 Aug 22
1111 Oct 04
1125 Dec 26
1117 Jan 04
1102 Oct 13
1144 Jan 06
1131 Mar 30
1130 Apr 09
1105 Aug 11
1140 Mar 20

Total Eclipse	
Annular Eclipse	
Greatest Eclipse	✳

0   500   1000
Kilometers
©2015 by Fred Espenak, Astropixels.com

Map 04: Central Solar Eclipses of 1151–1200 CE

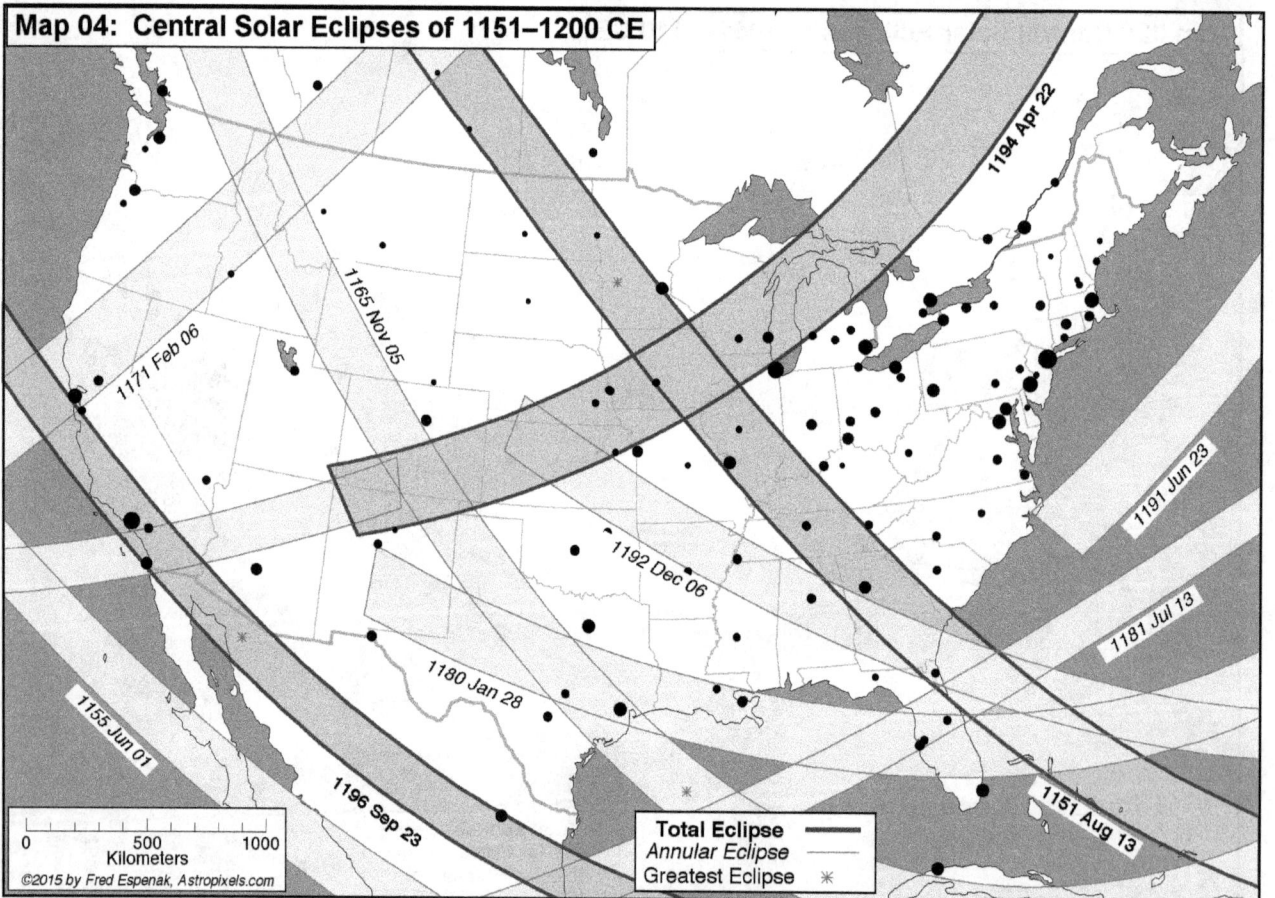

1194 Apr 22
1165 Nov 05
1171 Feb 06
1191 Jun 23
1192 Dec 06
1181 Jul 13
1180 Jan 28
1155 Jun 01
1196 Sep 23
1151 Aug 13

Total Eclipse	
Annular Eclipse	
Greatest Eclipse	

0   500   1000
Kilometers
©2015 by Fred Espenak, Astropixels.com

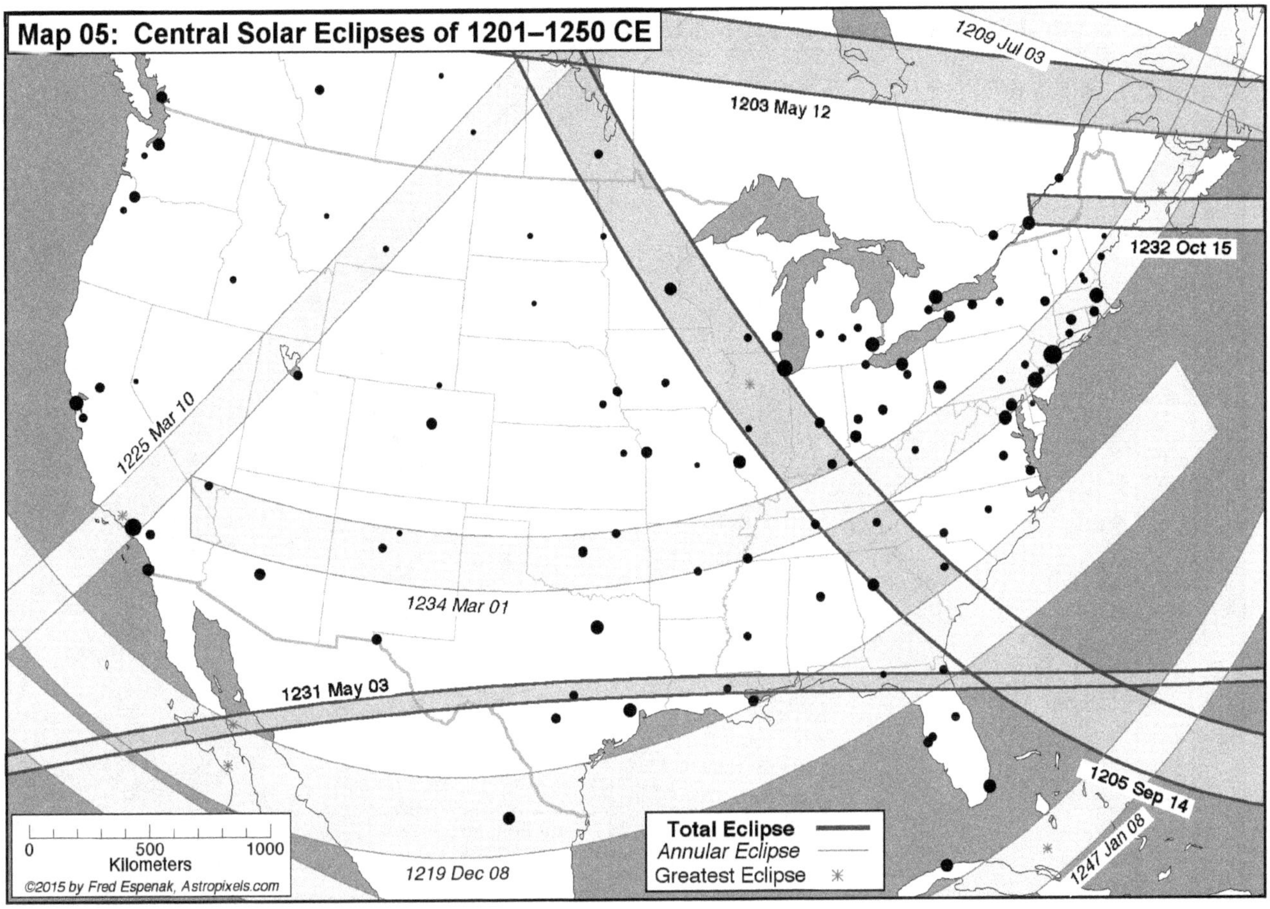

Map 05: Central Solar Eclipses of 1201–1250 CE

1209 Jul 03
1203 May 12
1232 Oct 15
1225 Mar 10
1234 Mar 01
1231 May 03
1205 Sep 14
1247 Jan 08
1219 Dec 08

Total Eclipse
Annular Eclipse
Greatest Eclipse ✳

0 500 1000
Kilometers
©2015 by Fred Espenak, Astropixels.com

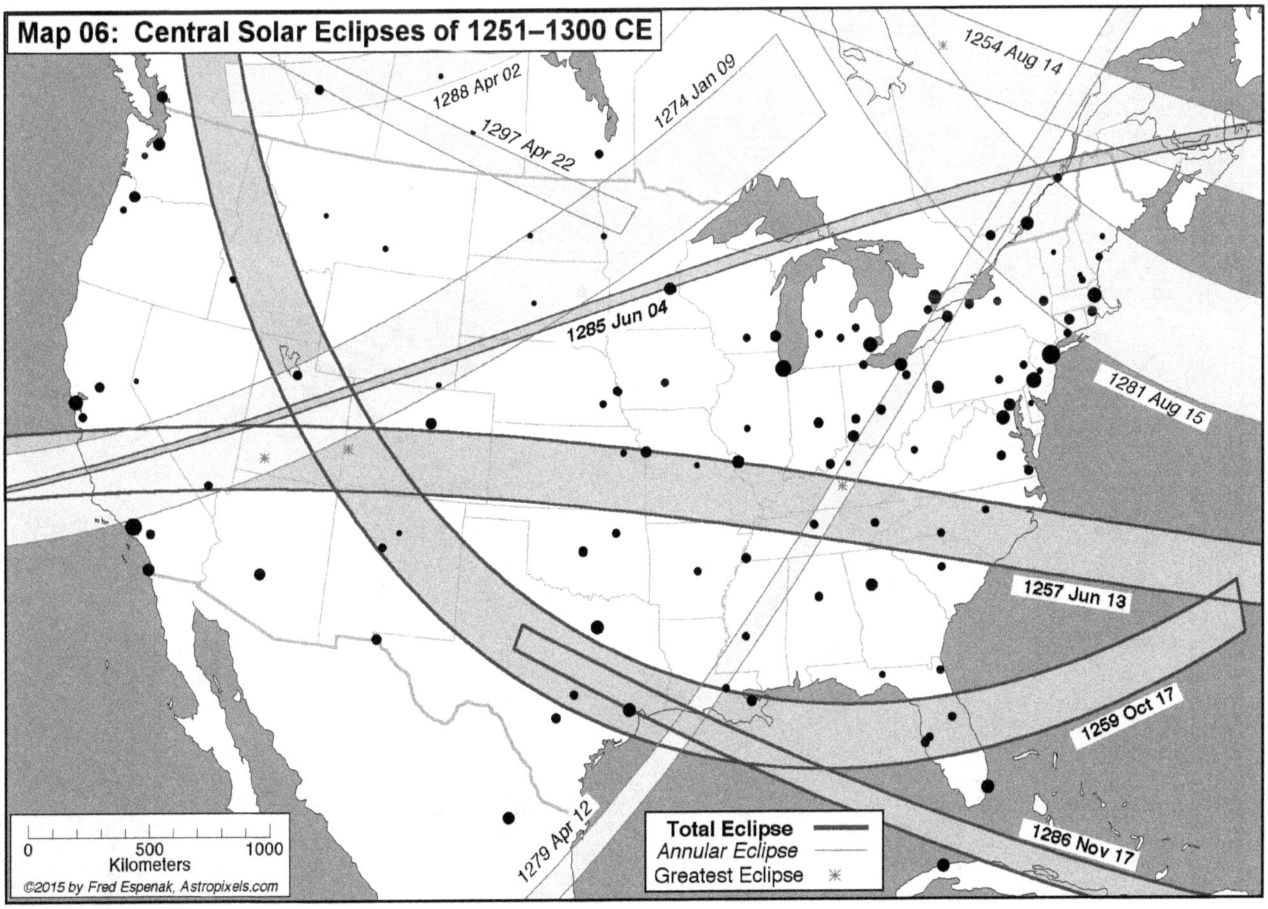

Map 06: Central Solar Eclipses of 1251–1300 CE

1254 Aug 14
1288 Apr 02
1274 Jan 09
1297 Apr 22
1285 Jun 04
1281 Aug 15
1257 Jun 13
1259 Oct 17
1279 Apr 12
1286 Nov 17

Total Eclipse
Annular Eclipse
Greatest Eclipse ✳

0 500 1000
Kilometers
©2015 by Fred Espenak, Astropixels.com

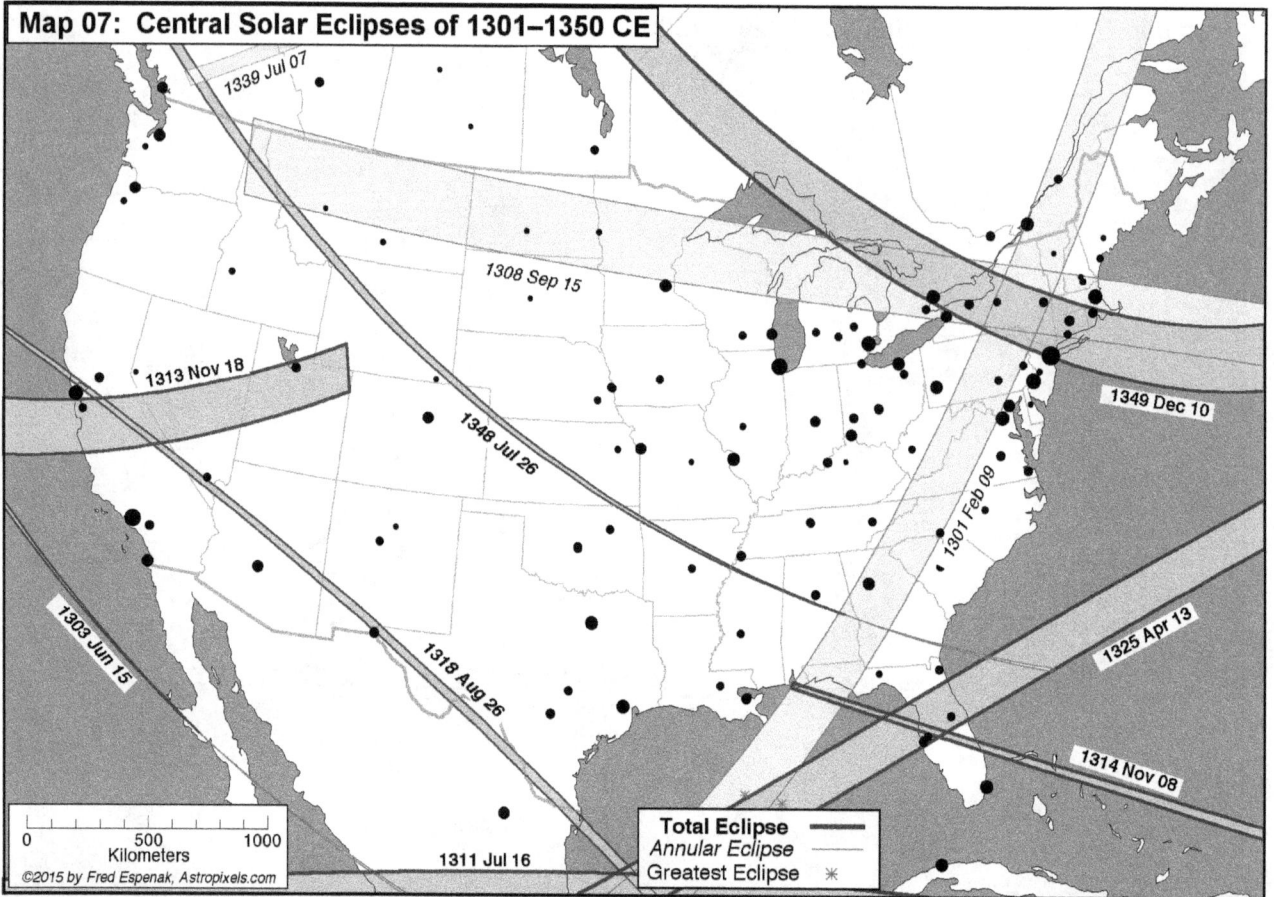

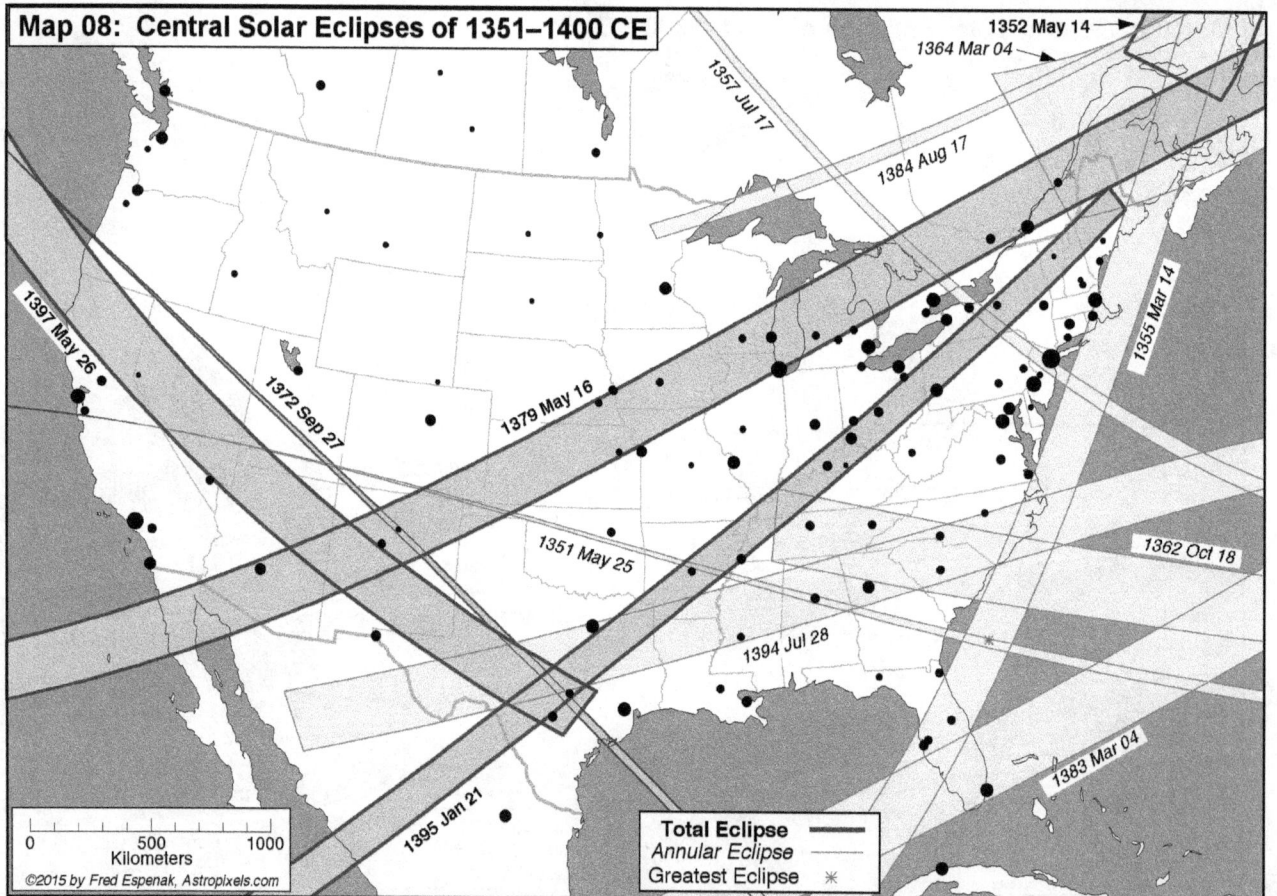

Map 07: Central Solar Eclipses of 1301–1350 CE

1339 Jul 07
1308 Sep 15
1313 Nov 18
1348 Jul 26
1349 Dec 10
1301 Feb 09
1303 Jun 15
1325 Apr 13
1318 Aug 26
1314 Nov 08
1311 Jul 16

Total Eclipse
Annular Eclipse
Greatest Eclipse ✳

0 500 1000
Kilometers
©2015 by Fred Espenak, Astropixels.com

Map 08: Central Solar Eclipses of 1351–1400 CE

1352 May 14
1364 Mar 04
1357 Jul 17
1384 Aug 17
1355 Mar 14
1397 May 26
1372 Sep 27
1379 May 16
1362 Oct 18
1351 May 25
1394 Jul 28
1383 Mar 04
1395 Jan 21

Total Eclipse
Annular Eclipse
Greatest Eclipse ✳

0 500 1000
Kilometers
©2015 by Fred Espenak, Astropixels.com

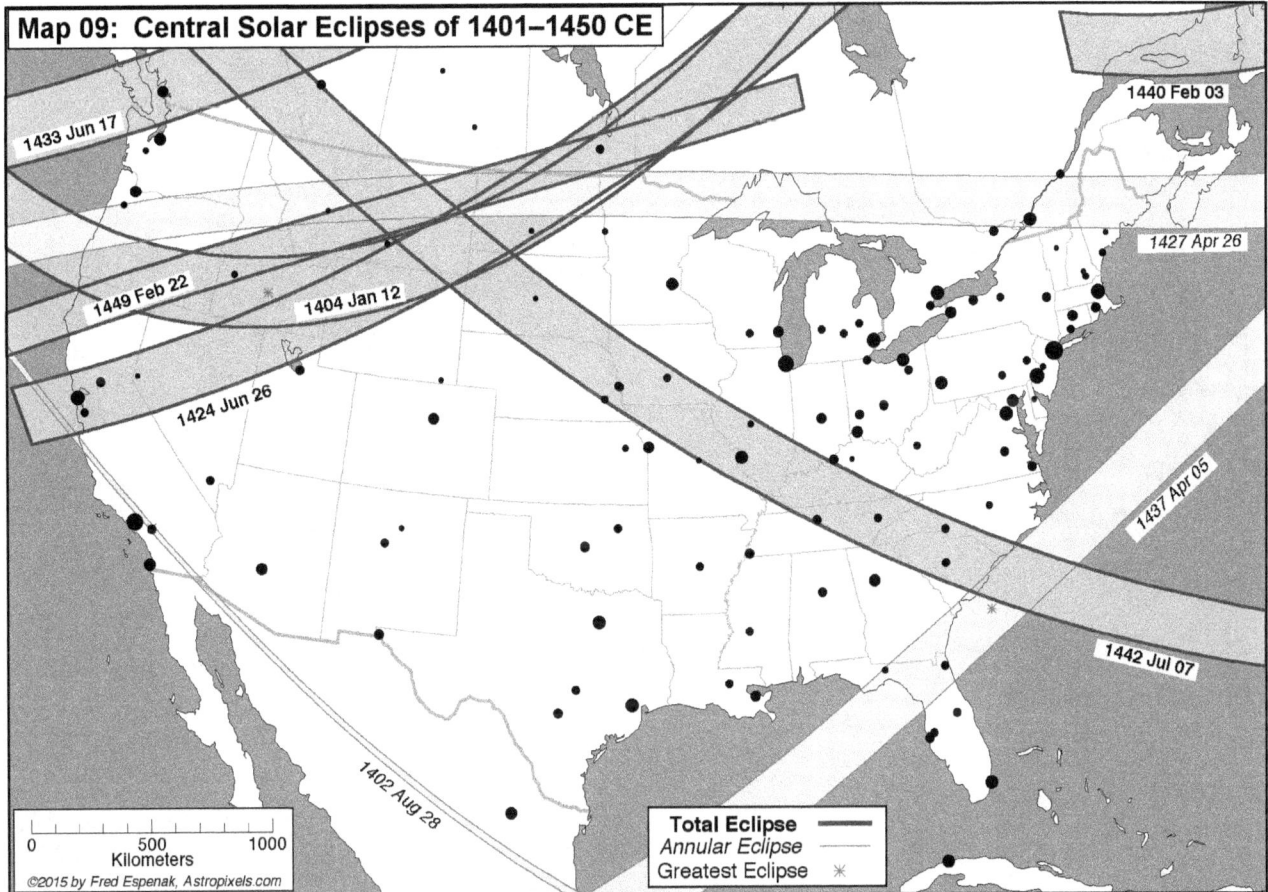

Map 09: Central Solar Eclipses of 1401–1450 CE

1433 Jun 17
1440 Feb 03
1427 Apr 26
1449 Feb 22
1404 Jan 12
1424 Jun 26
1437 Apr 05
1442 Jul 07
1402 Aug 28

Total Eclipse	——
Annular Eclipse	——
Greatest Eclipse	✳

0 500 1000
Kilometers
©2015 by Fred Espenak, Astropixels.com

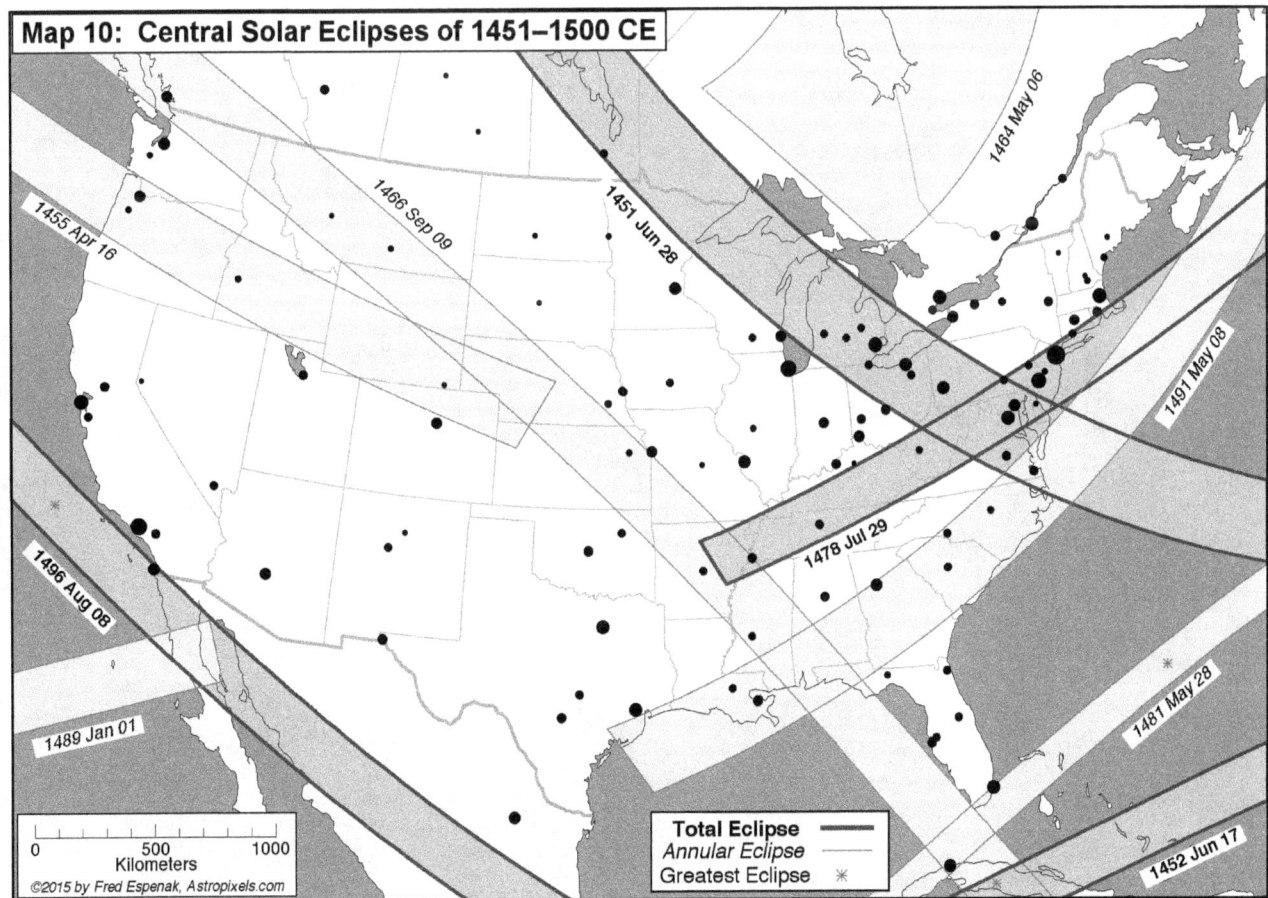

Map 10: Central Solar Eclipses of 1451–1500 CE

1464 May 06
1455 Apr 16
1466 Sep 09
1451 Jun 28
1491 May 08
1478 Jul 29
1496 Aug 06
1481 May 28
1489 Jan 01
1452 Jun 17

Total Eclipse	——
Annular Eclipse	——
Greatest Eclipse	✳

0 500 1000
Kilometers
©2015 by Fred Espenak, Astropixels.com

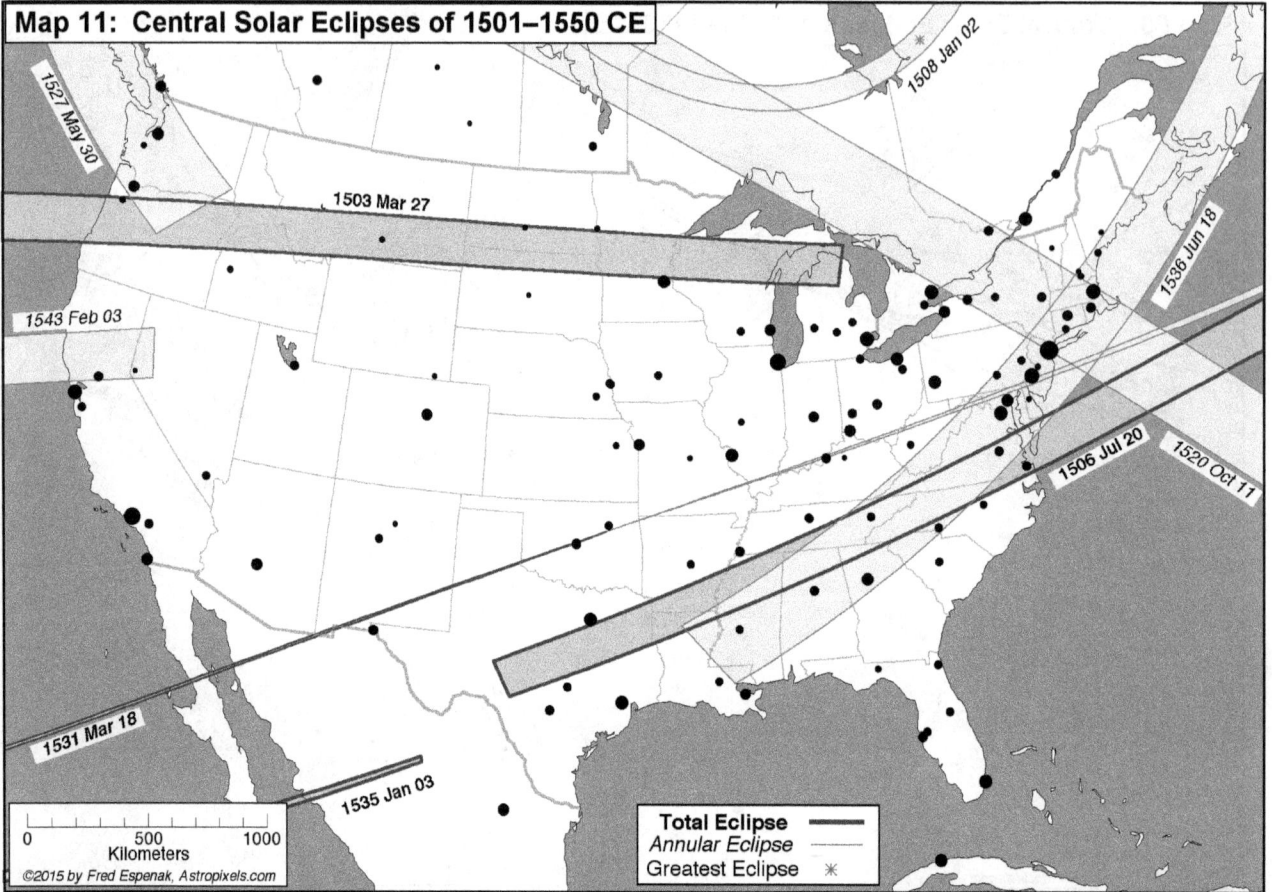

Map 11: Central Solar Eclipses of 1501–1550 CE

1508 Jan 02
1527 May 30
1503 Mar 27
1536 Jun 18
1543 Feb 03
1506 Jul 20
1520 Oct 11
1531 Mar 18
1535 Jan 03

Total Eclipse	——
Annular Eclipse	
Greatest Eclipse	✳

0 500 1000
Kilometers
©2015 by Fred Espenak, Astropixels.com

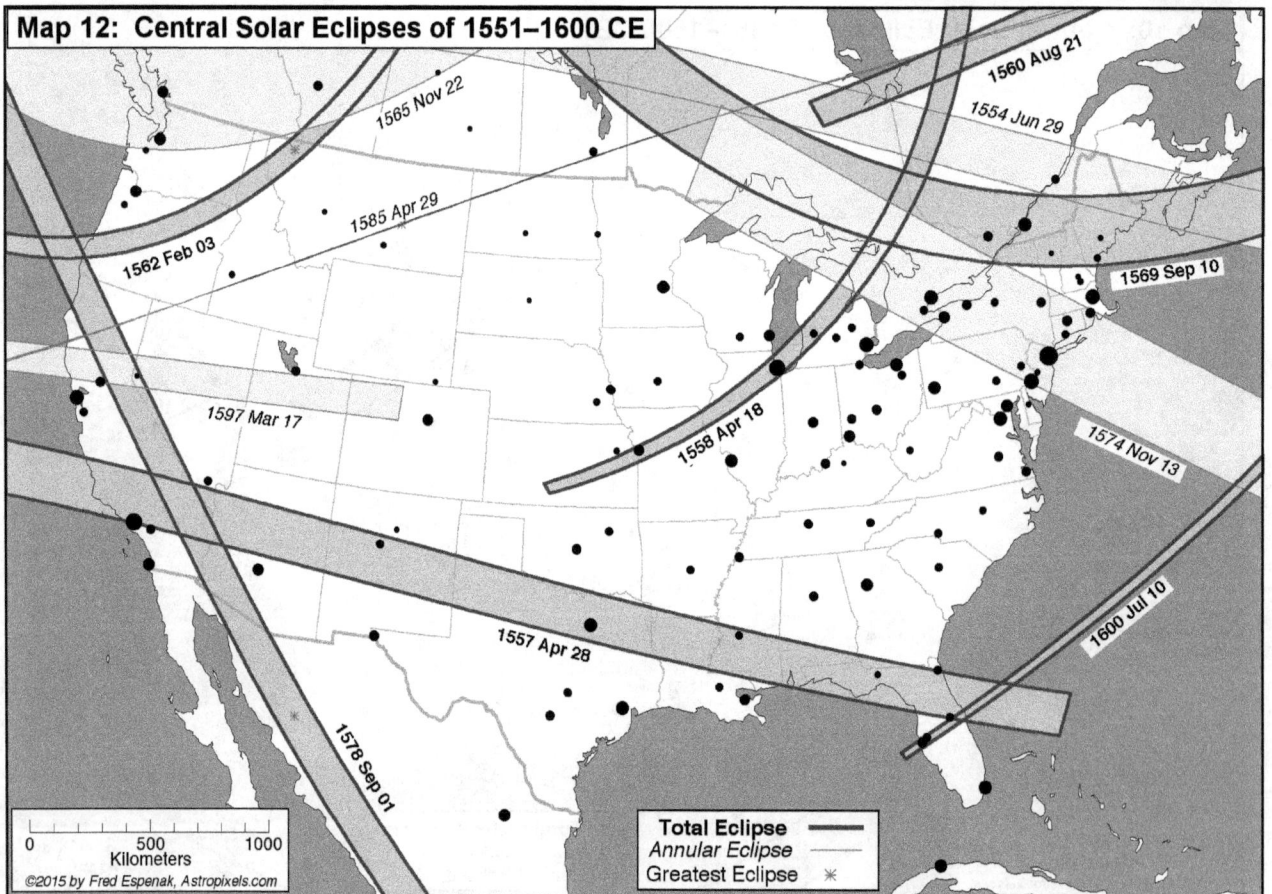

Map 12: Central Solar Eclipses of 1551–1600 CE

1560 Aug 21
1565 Nov 22
1554 Jun 29
1585 Apr 29
1562 Feb 03
1569 Sep 10
1597 Mar 17
1558 Apr 18
1574 Nov 13
1557 Apr 28
1600 Jul 10
1578 Sep 01

Total Eclipse	——
Annular Eclipse	
Greatest Eclipse	✳

0 500 1000
Kilometers
©2015 by Fred Espenak, Astropixels.com

Map 13: Central Solar Eclipses of 1601–1650 CE

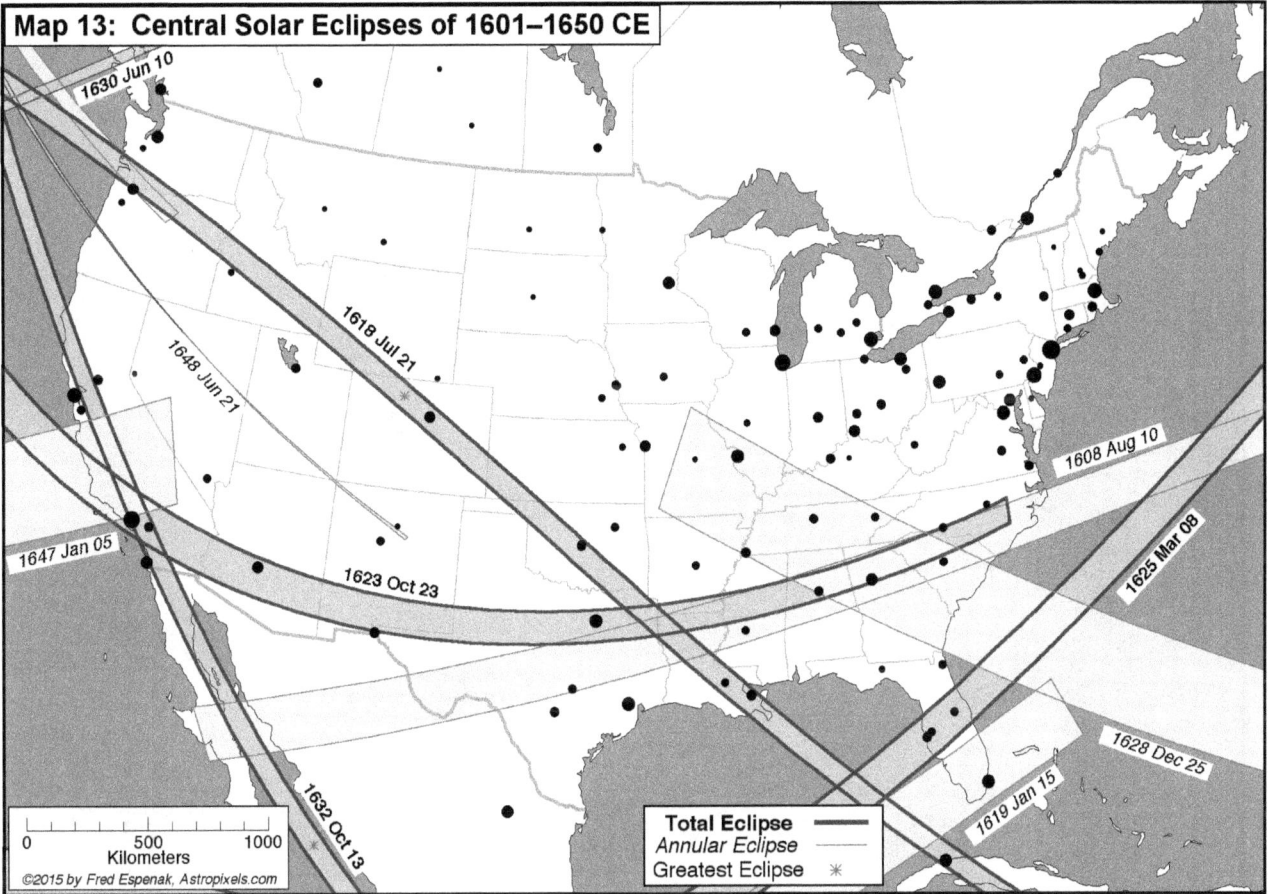

1630 Jun 10
1618 Jul 21
1648 Jun 21
1647 Jan 05
1623 Oct 23
1632 Oct 13
1608 Aug 10
1625 Mar 08
1628 Dec 25
1619 Jan 15

0	500	1000
Kilometers		

©2015 by Fred Espenak, Astropixels.com

Total Eclipse	——
Annular Eclipse	——
Greatest Eclipse	✳

Map 14: Central Solar Eclipses of 1651–1700 CE

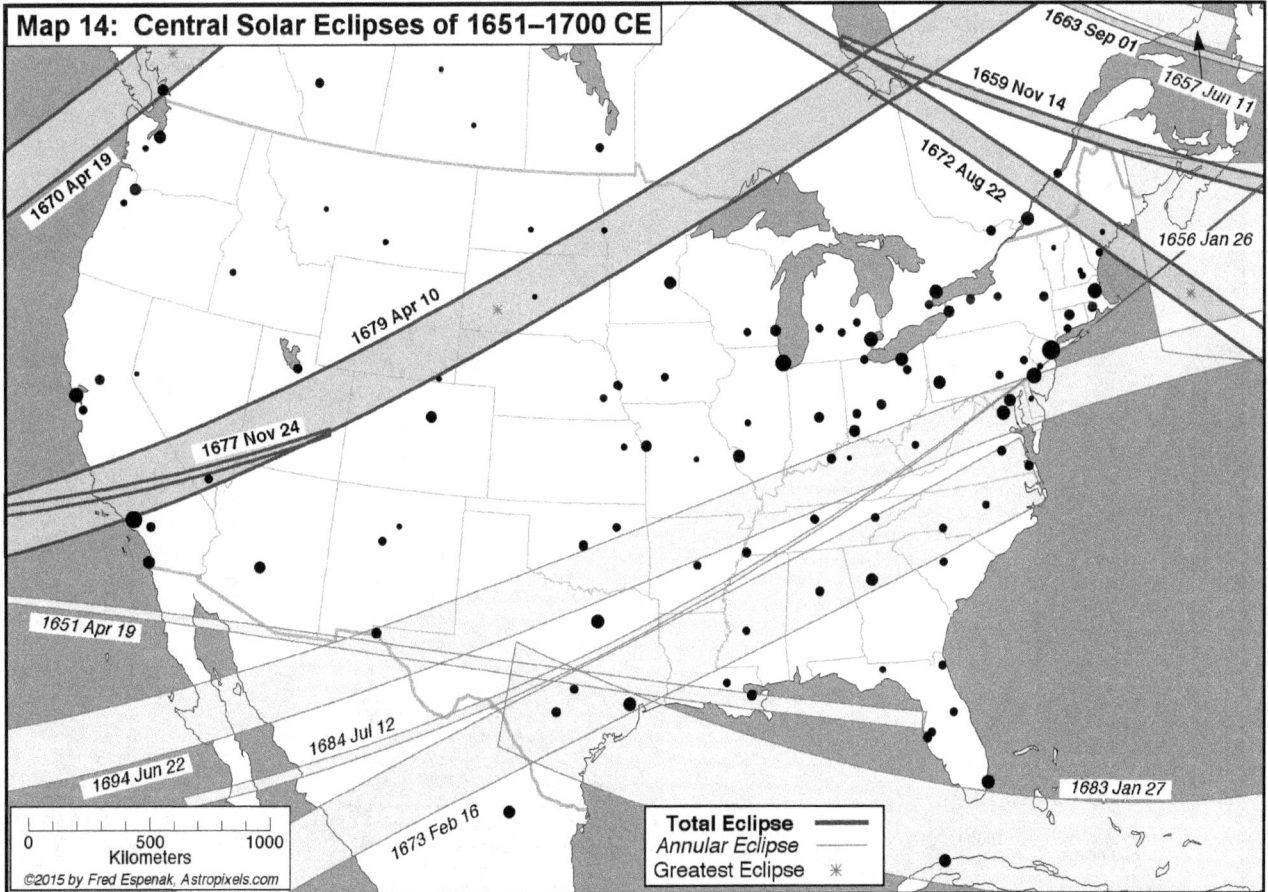

1663 Sep 01
1659 Nov 14
1657 Jun 11
1670 Apr 19
1672 Aug 22
1656 Jan 26
1679 Apr 10
1677 Nov 24
1651 Apr 19
1684 Jul 12
1694 Jun 22
1683 Jan 27
1673 Feb 16

0	500	1000
Kilometers		

©2015 by Fred Espenak, Astropixels.com

Total Eclipse	——
Annular Eclipse	——
Greatest Eclipse	✳

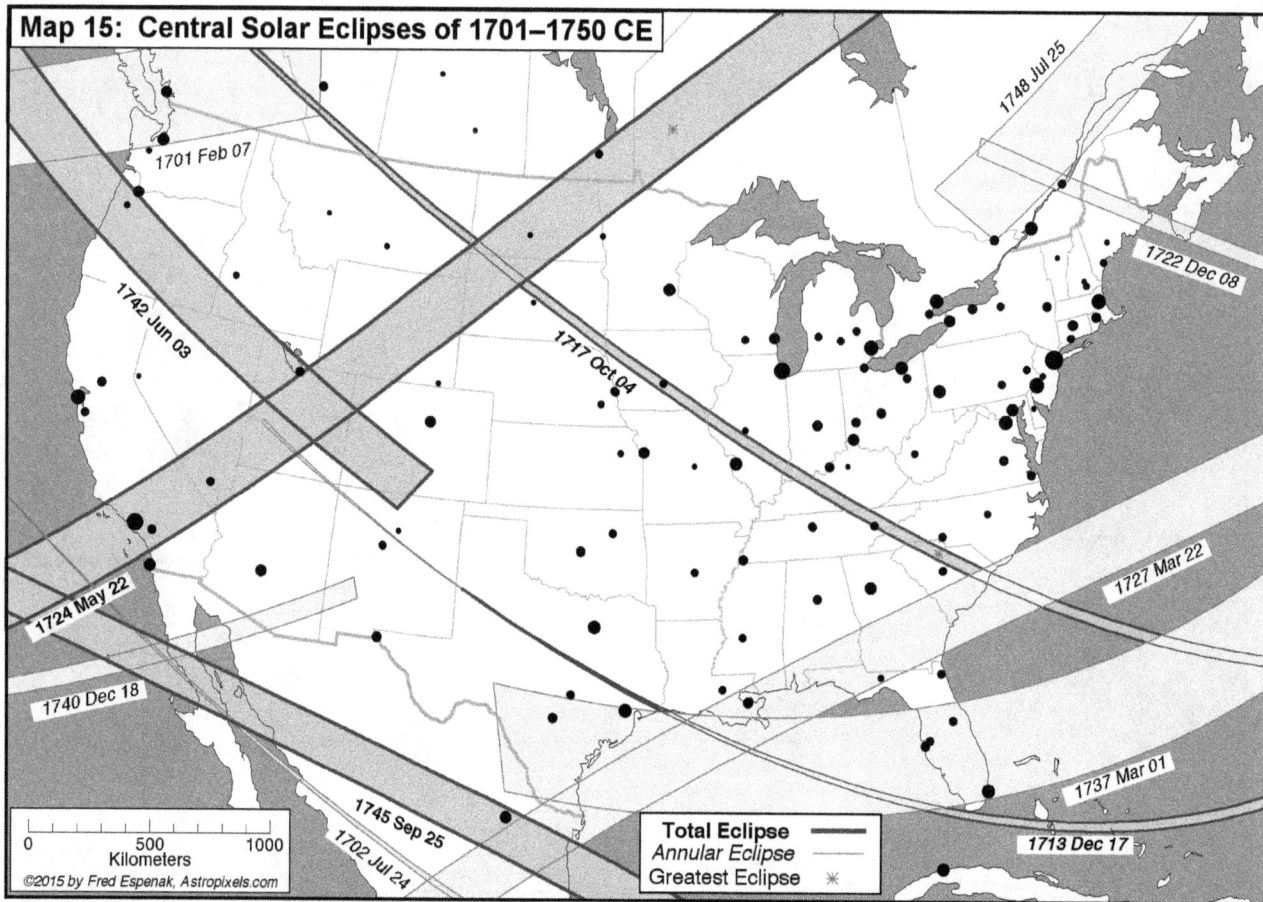

Map 15: Central Solar Eclipses of 1701–1750 CE

1748 Jul 25
1701 Feb 07
1742 Jun 03
1717 Oct 04
1722 Dec 08
1727 Mar 22
1724 May 22
1740 Dec 18
1737 Mar 01
1745 Sep 25
1702 Jul 24
1713 Dec 17

Total Eclipse	———
Annular Eclipse	——
Greatest Eclipse	✳

0   500   1000
Kilometers
©2015 by Fred Espenak, Astropixels.com

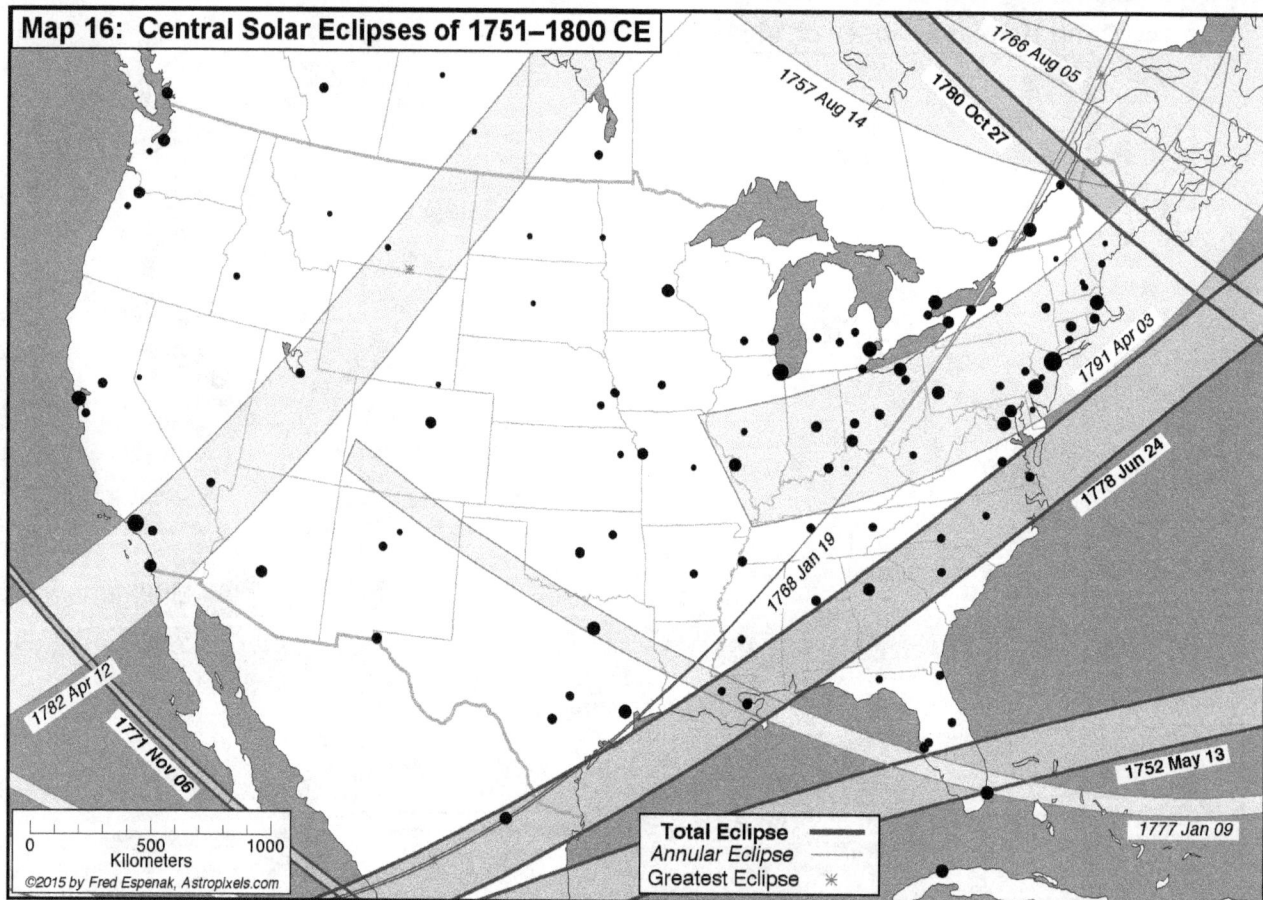

Map 16: Central Solar Eclipses of 1751–1800 CE

1766 Aug 05
1757 Aug 14
1780 Oct 27
1791 Apr 03
1778 Jun 24
1768 Jan 19
1782 Apr 12
1771 Nov 06
1752 May 13
1777 Jan 09

Total Eclipse	———
Annular Eclipse	——
Greatest Eclipse	✳

0   500   1000
Kilometers
©2015 by Fred Espenak, Astropixels.com

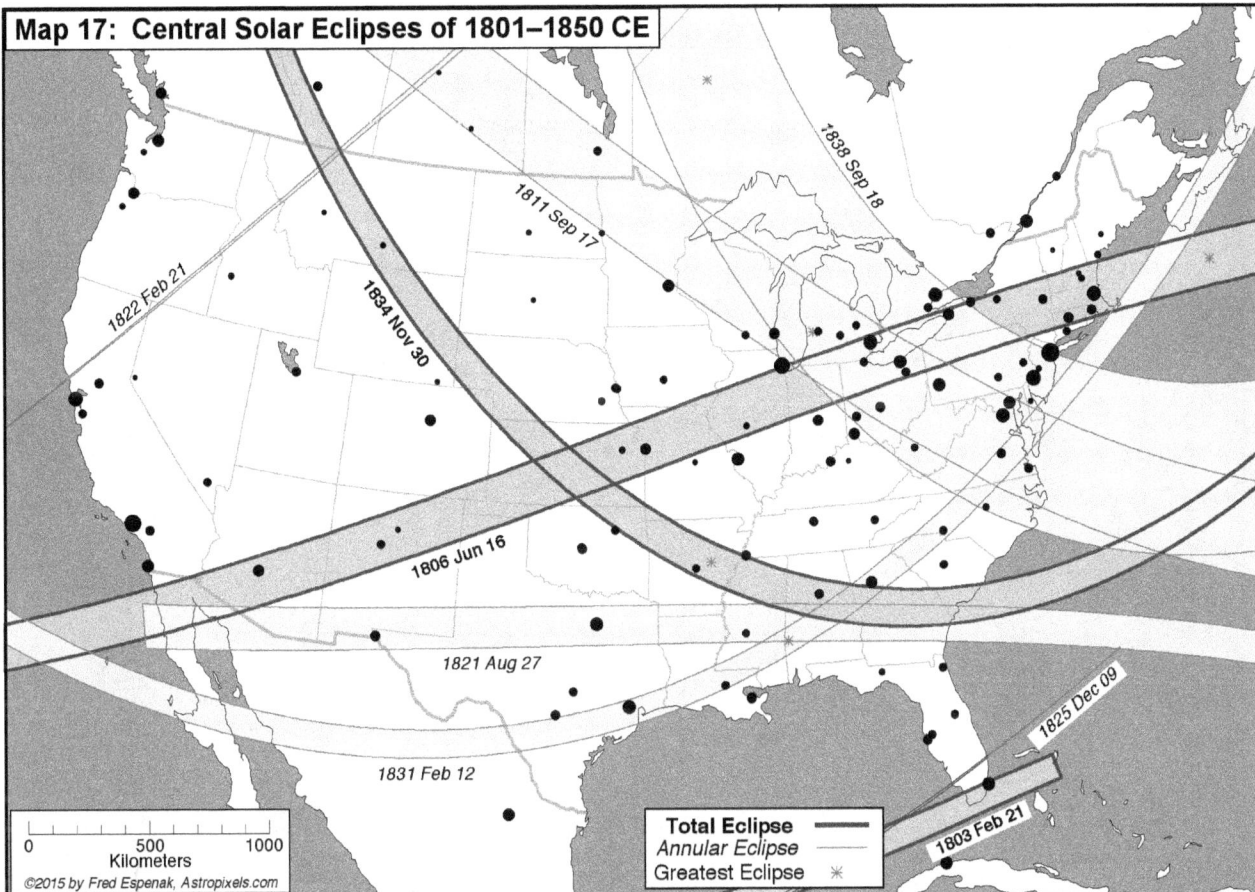

Map 17: Central Solar Eclipses of 1801–1850 CE

1838 Sep 18

1811 Sep 17

1822 Feb 21

1834 Nov 30

1806 Jun 16

1821 Aug 27

1831 Feb 12

1825 Dec 09

1803 Feb 21

Total Eclipse
Annular Eclipse
Greatest Eclipse ✳

0   500   1000
Kilometers
©2015 by Fred Espenak, Astropixels.com

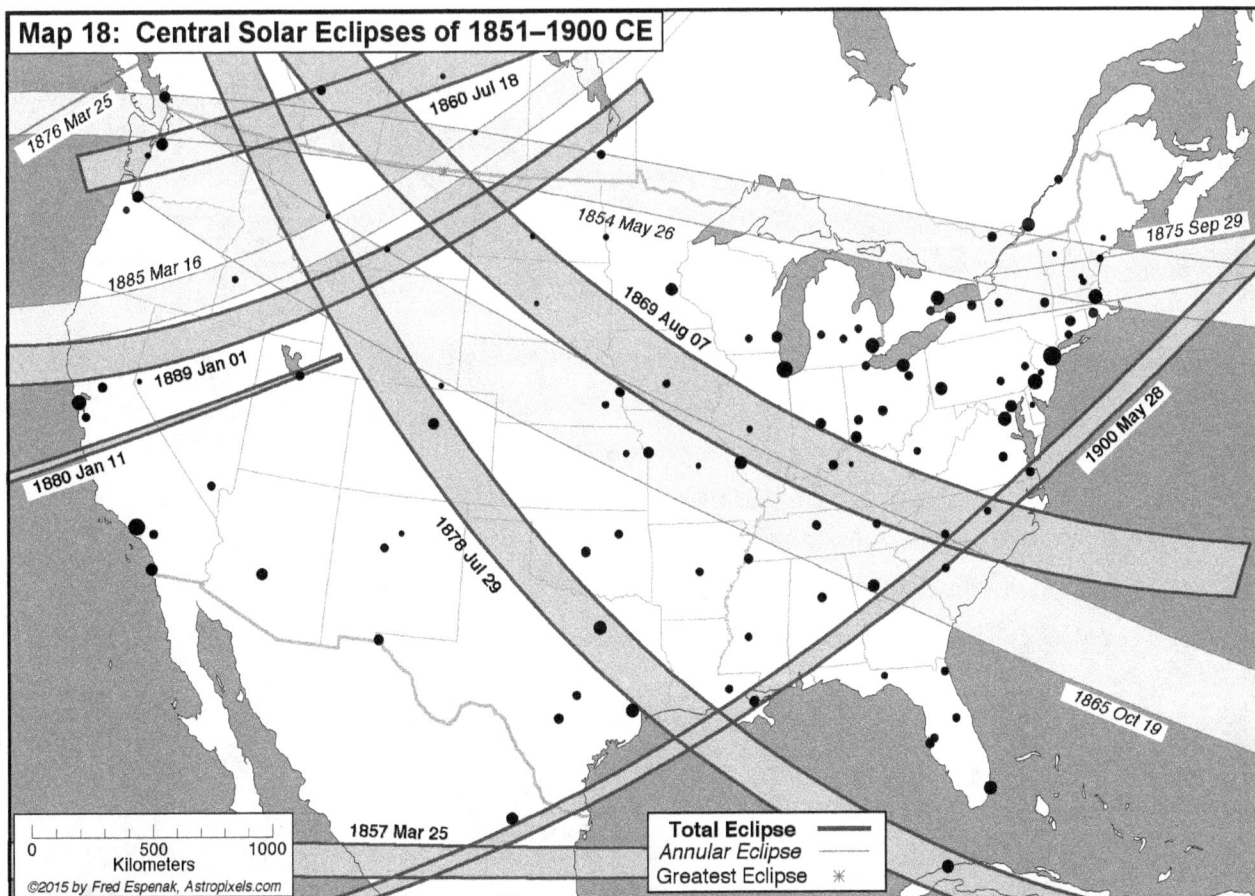

Map 18: Central Solar Eclipses of 1851–1900 CE

1876 Mar 25

1860 Jul 18

1854 May 26

1875 Sep 29

1885 Mar 16

1869 Aug 07

1889 Jan 01

1900 May 28

1880 Jan 11

1878 Jul 29

1865 Oct 19

1857 Mar 25

Total Eclipse
Annular Eclipse
Greatest Eclipse ✳

0   500   1000
Kilometers
©2015 by Fred Espenak, Astropixels.com

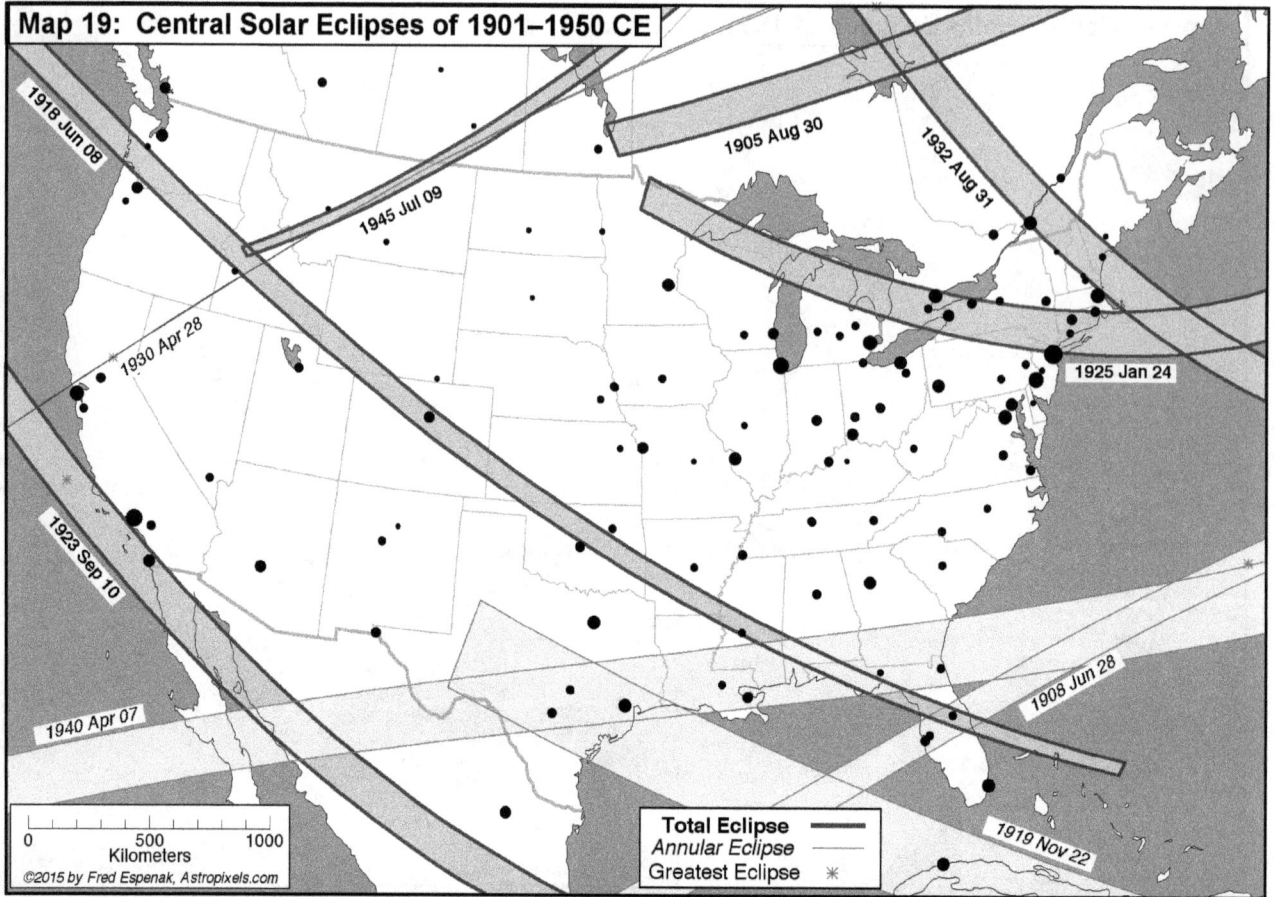

Map 19: Central Solar Eclipses of 1901–1950 CE

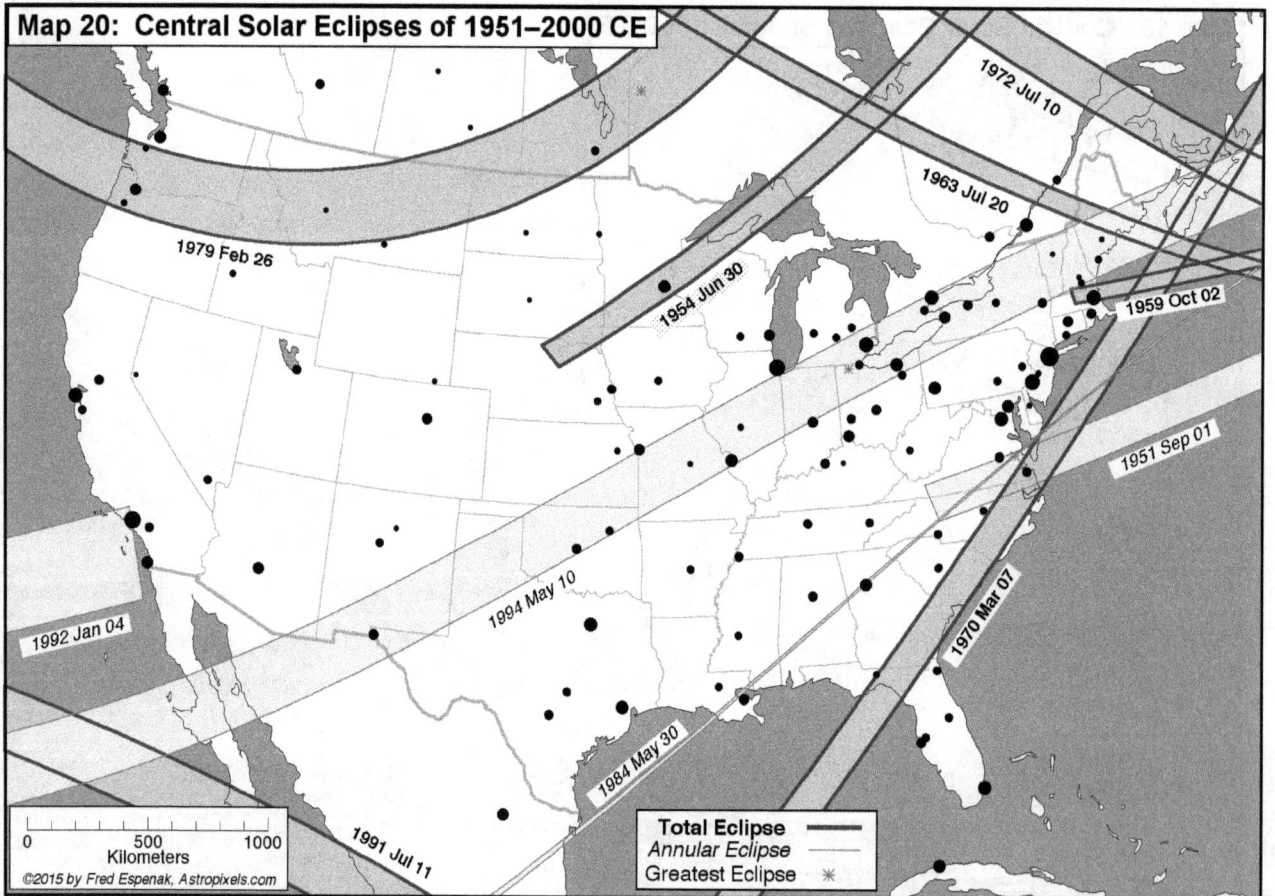

Map 20: Central Solar Eclipses of 1951–2000 CE

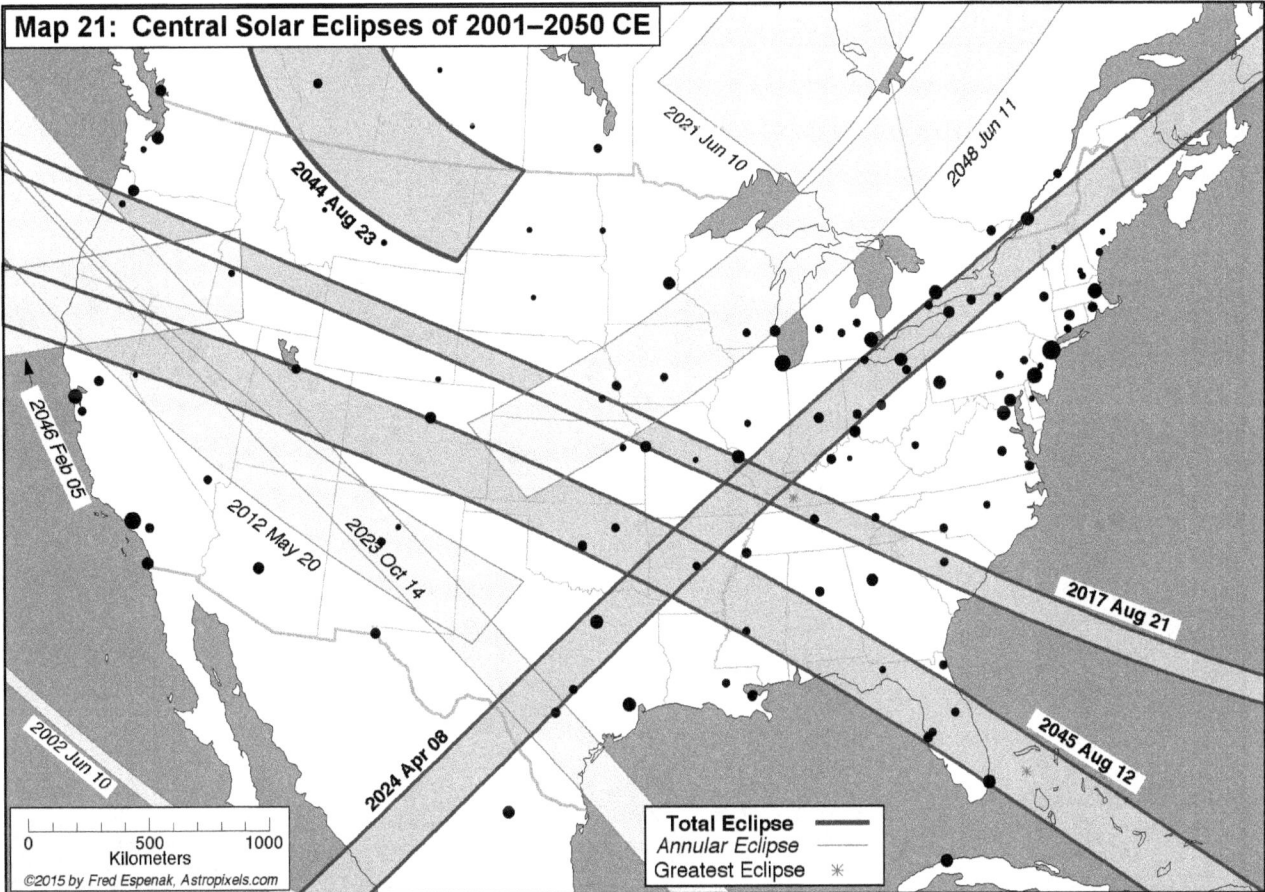

Map 21: Central Solar Eclipses of 2001–2050 CE

2021 Jun 10
2048 Jun 11
2044 Aug 23
2046 Feb 05
2012 May 20
2023 Oct 14
2017 Aug 21
2002 Jun 10
2024 Apr 08
2045 Aug 12

Kilometers
0 500 1000

©2015 by Fred Espenak, Astropixels.com

Total Eclipse
Annular Eclipse
Greatest Eclipse   *

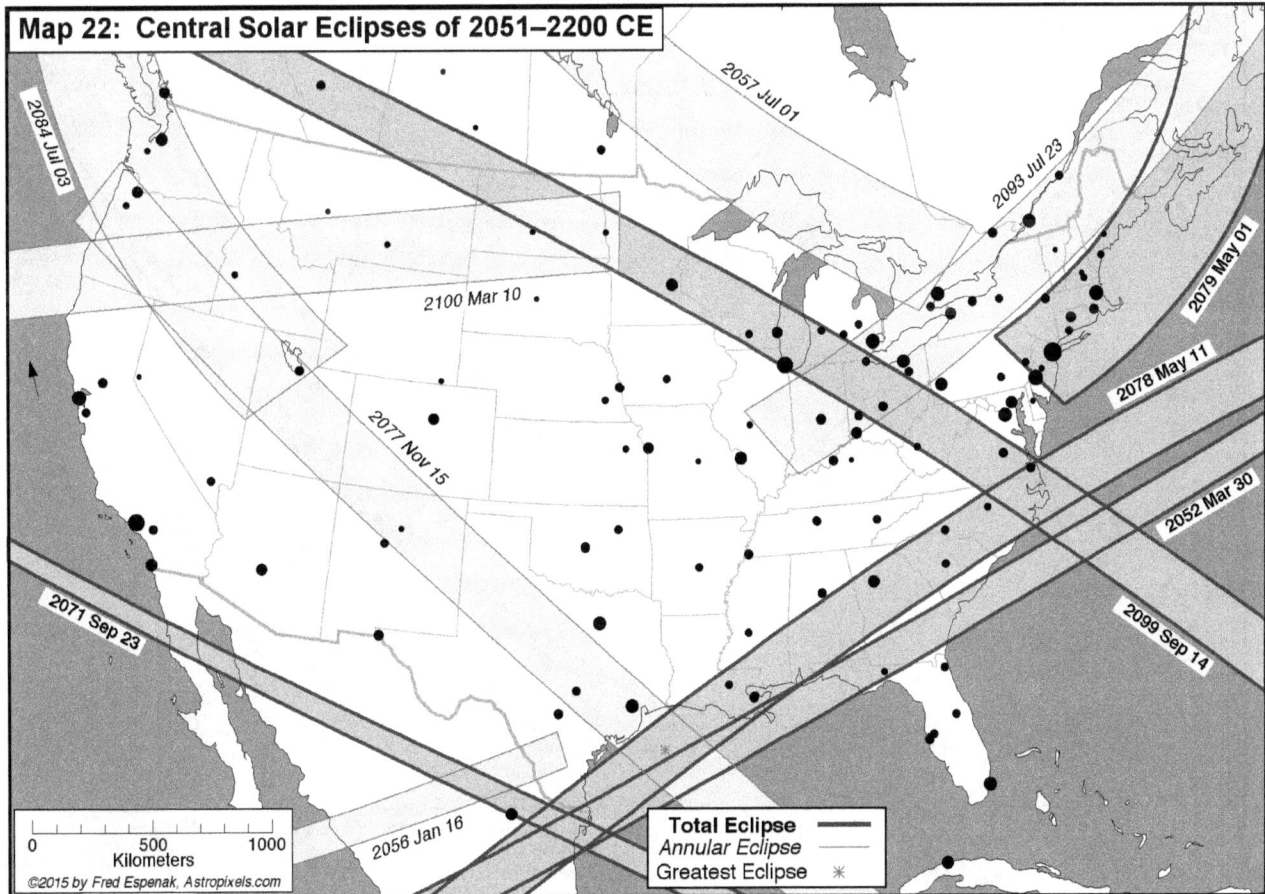

Map 22: Central Solar Eclipses of 2051–2200 CE

2057 Jul 01
2084 Jul 03
2093 Jul 23
2079 May 01
2100 Mar 10
2078 May 11
2077 Nov 15
2052 Mar 30
2071 Sep 23
2099 Sep 14
2056 Jan 16

Kilometers
0 500 1000

©2015 by Fred Espenak, Astropixels.com

Total Eclipse
Annular Eclipse
Greatest Eclipse   *

Map 23: Central Solar Eclipses of 2101–2150 CE

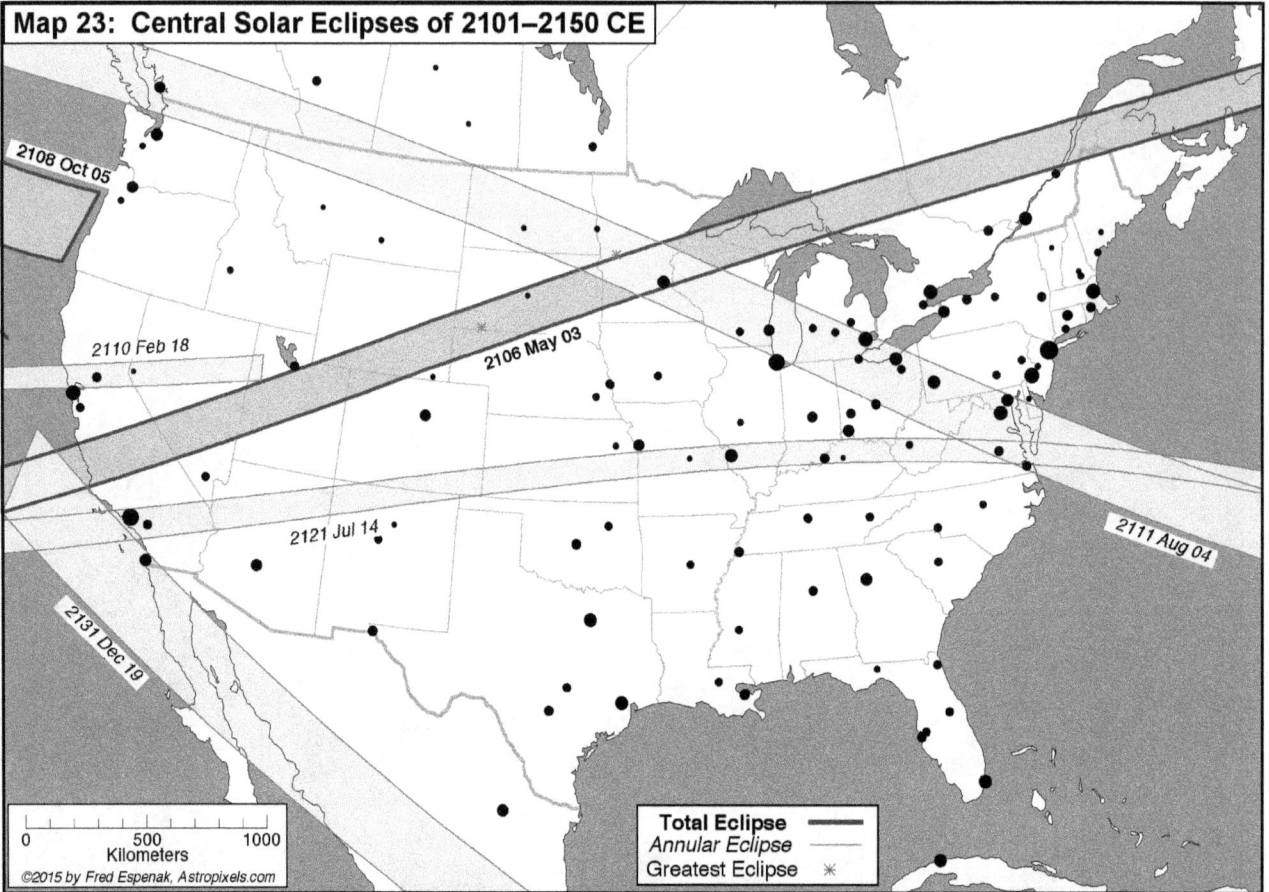

2108 Oct 05
2110 Feb 18
2106 May 03
2121 Jul 14
2131 Dec 19
2111 Aug 04

Total Eclipse	——
Annular Eclipse	——
Greatest Eclipse	✳

0 500 1000
Kilometers
©2015 by Fred Espenak, Astropixels.com

Map 24: Central Solar Eclipses of 2151–2200 CE

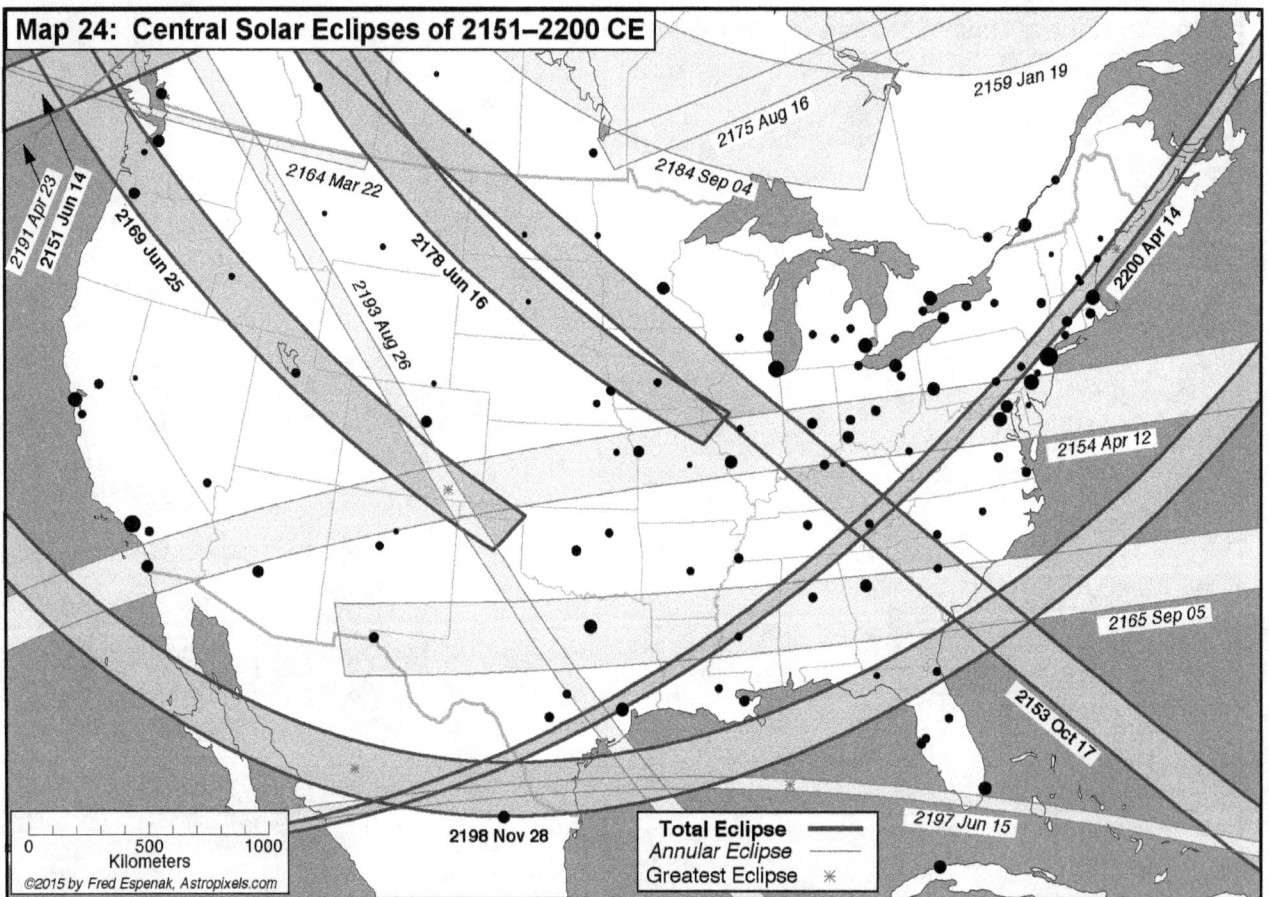

2159 Jan 19
2175 Aug 16
2184 Sep 04
2164 Mar 22
2178 Jun 16
2191 Apr 23
2151 Jun 14
2169 Jun 25
2193 Aug 26
2200 Apr 14
2154 Apr 12
2165 Sep 05
2153 Oct 17
2197 Jun 15
2198 Nov 28

Total Eclipse	——
Annular Eclipse	——
Greatest Eclipse	✳

0 500 1000
Kilometers
©2015 by Fred Espenak, Astropixels.com

Map 25: Central Solar Eclipses of 2201–2250 CE

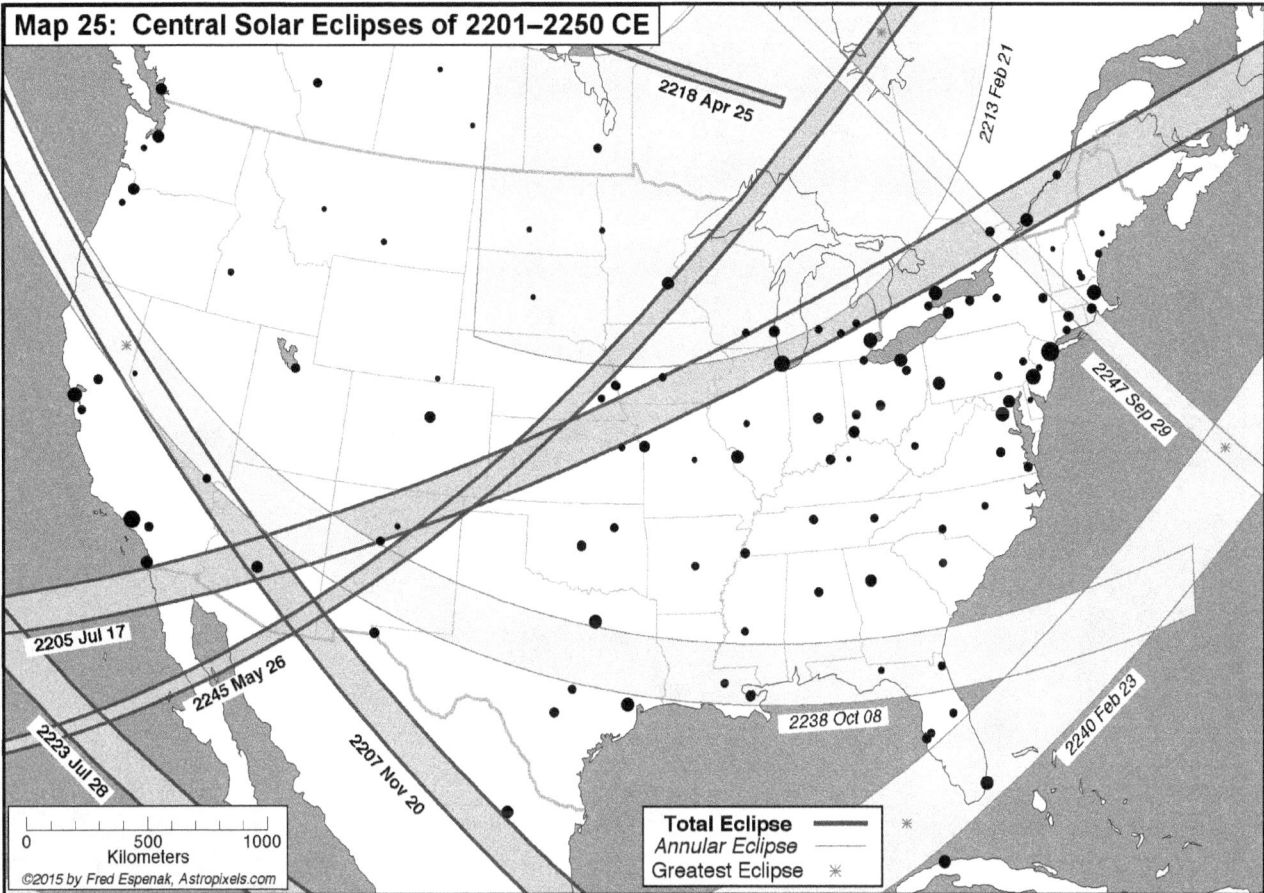

2218 Apr 25

2213 Feb 21

2247 Sep 29

2205 Jul 17

2245 May 26

2223 Jul 28

2207 Nov 20

2238 Oct 08

2240 Feb 23

Total Eclipse	———	✳
Annular Eclipse	———	
Greatest Eclipse	✳	

0   500   1000
Kilometers

©2015 by Fred Espenak, Astropixels.com

Map 26: Central Solar Eclipses of 2251–2300 CE

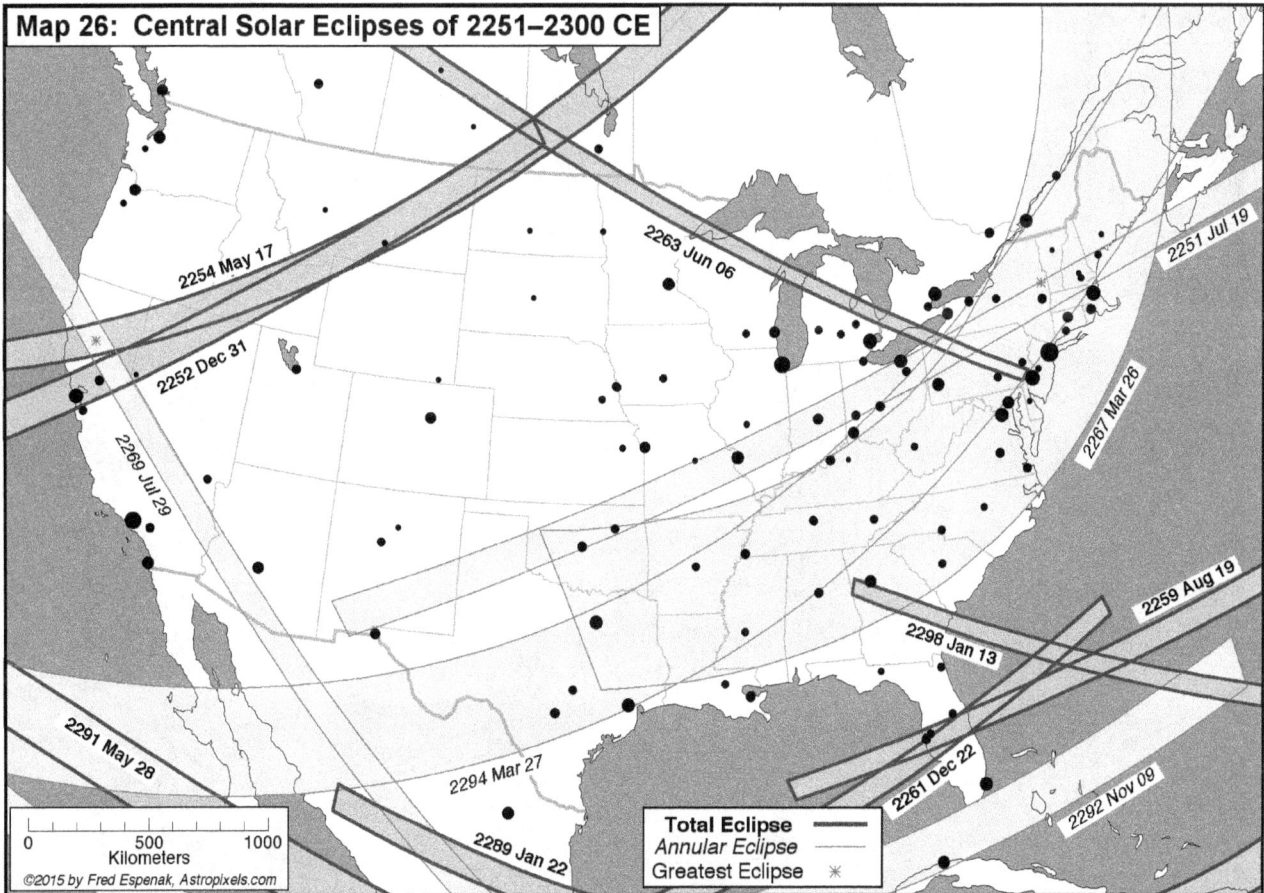

2263 Jun 06

2251 Jul 19

2254 May 17

2252 Dec 31

2267 Mar 26

2269 Jul 29

2259 Aug 19

2298 Jan 13

2291 May 28

2294 Mar 27

2261 Dec 22

2292 Nov 09

2289 Jan 22

Total Eclipse	———	
Annular Eclipse	———	
Greatest Eclipse	✳	

0   500   1000
Kilometers

©2015 by Fred Espenak, Astropixels.com

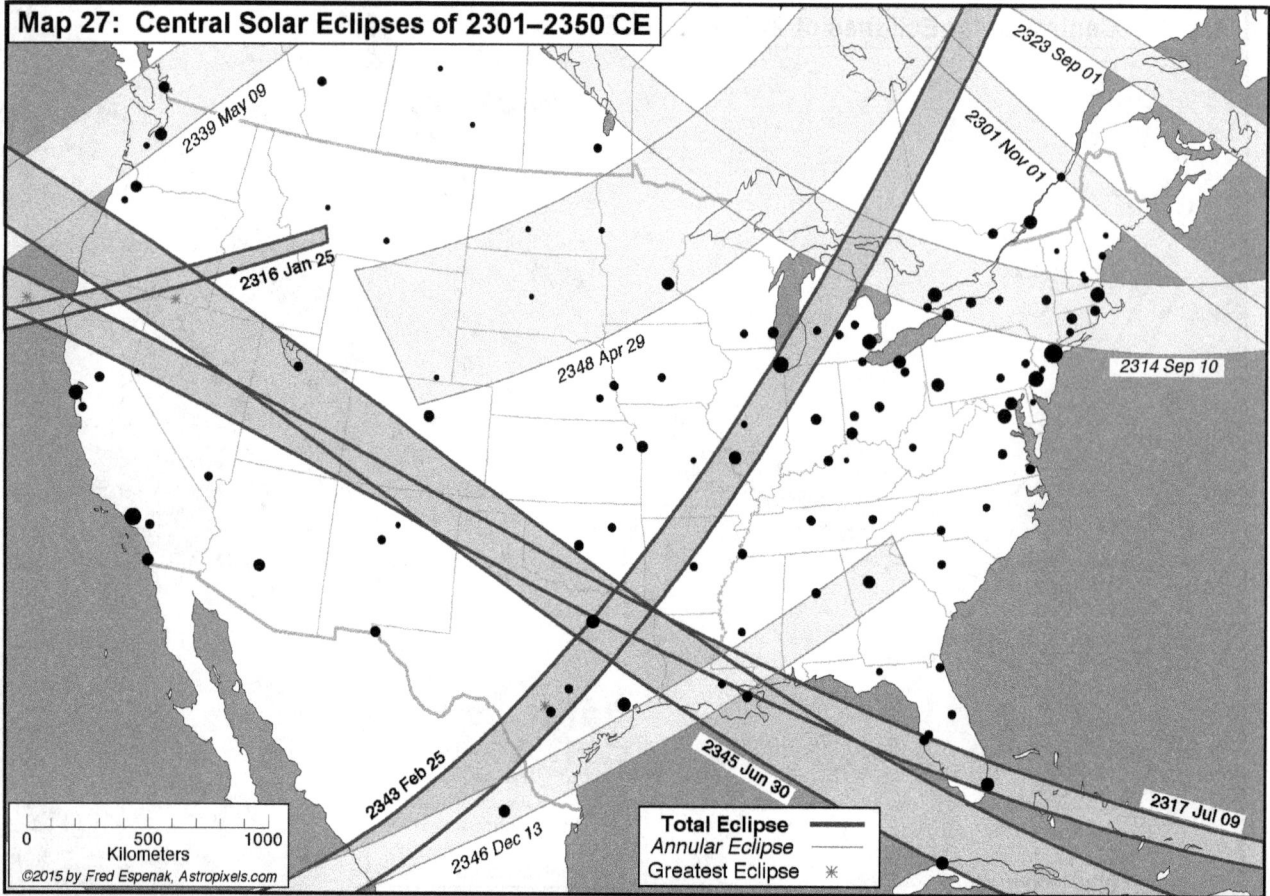

Map 27: Central Solar Eclipses of 2301–2350 CE

2323 Sep 01
2301 Nov 01
2339 May 09
2316 Jan 25
2348 Apr 29
2314 Sep 10
2343 Feb 25
2345 Jun 30
2346 Dec 13
2317 Jul 09

Total Eclipse	——
Annular Eclipse	——
Greatest Eclipse	✳

0   500   1000
Kilometers
©2015 by Fred Espenak, Astropixels.com

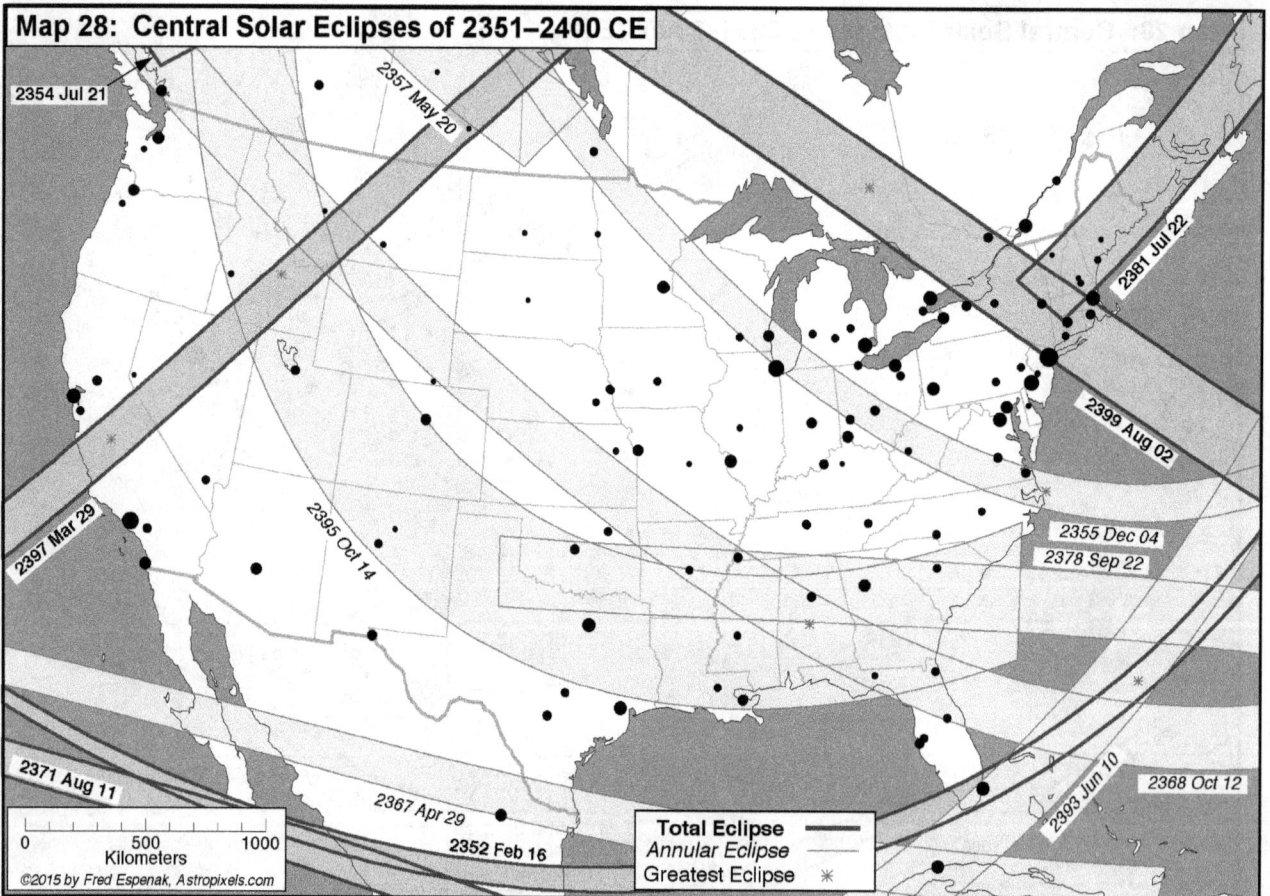

Map 28: Central Solar Eclipses of 2351–2400 CE

2354 Jul 21
2357 May 20
2381 Jul 22
2399 Aug 02
2355 Dec 04
2378 Sep 22
2395 Oct 14
2397 Mar 29
2371 Aug 11
2367 Apr 29
2352 Feb 16
2393 Jun 10
2368 Oct 12

Total Eclipse	——
Annular Eclipse	——
Greatest Eclipse	✳

0   500   1000
Kilometers
©2015 by Fred Espenak, Astropixels.com

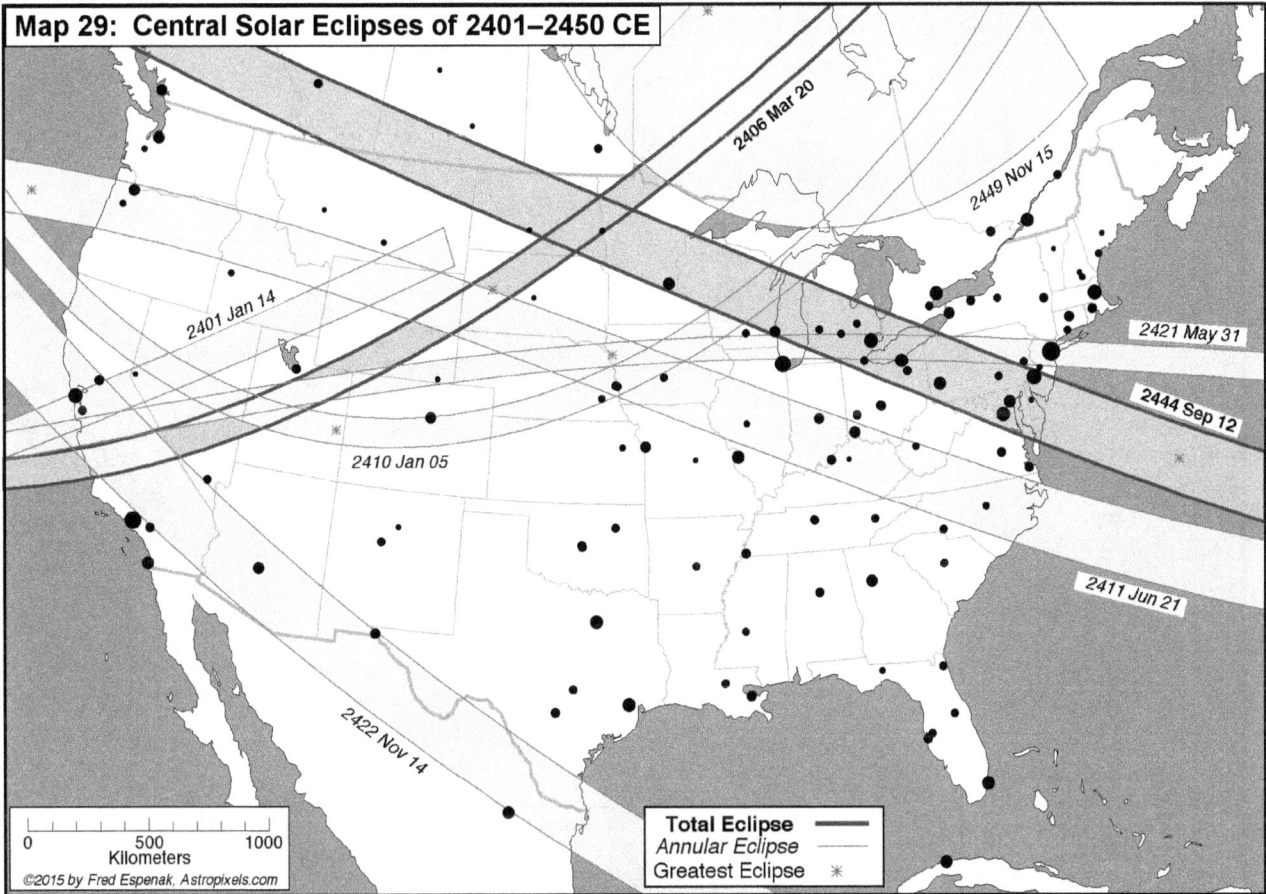

Map 29: Central Solar Eclipses of 2401–2450 CE

2406 Mar 20

2449 Nov 15

2401 Jan 14

2421 May 31

2444 Sep 12

2410 Jan 05

2411 Jun 21

2422 Nov 14

Total Eclipse	———
Annular Eclipse	—
Greatest Eclipse	*

0   500   1000
Kilometers

©2015 by Fred Espenak, Astropixels.com

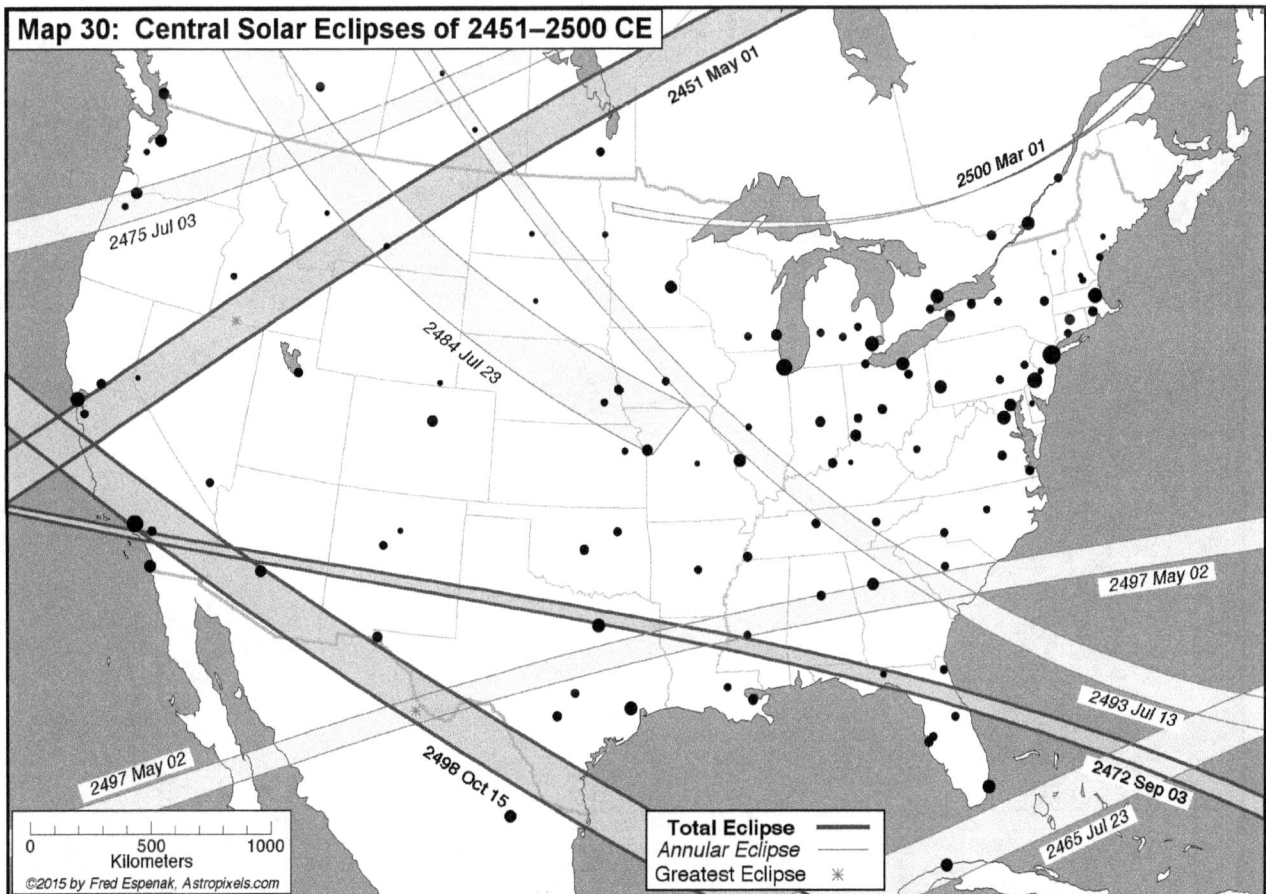

Map 30: Central Solar Eclipses of 2451–2500 CE

2451 May 01

2500 Mar 01

2475 Jul 03

2484 Jul 23

2497 May 02

2493 Jul 13

2497 May 02

2498 Oct 15

2472 Sep 03

2465 Jul 23

Total Eclipse	———
Annular Eclipse	—
Greatest Eclipse	*

0   500   1000
Kilometers

©2015 by Fred Espenak, Astropixels.com

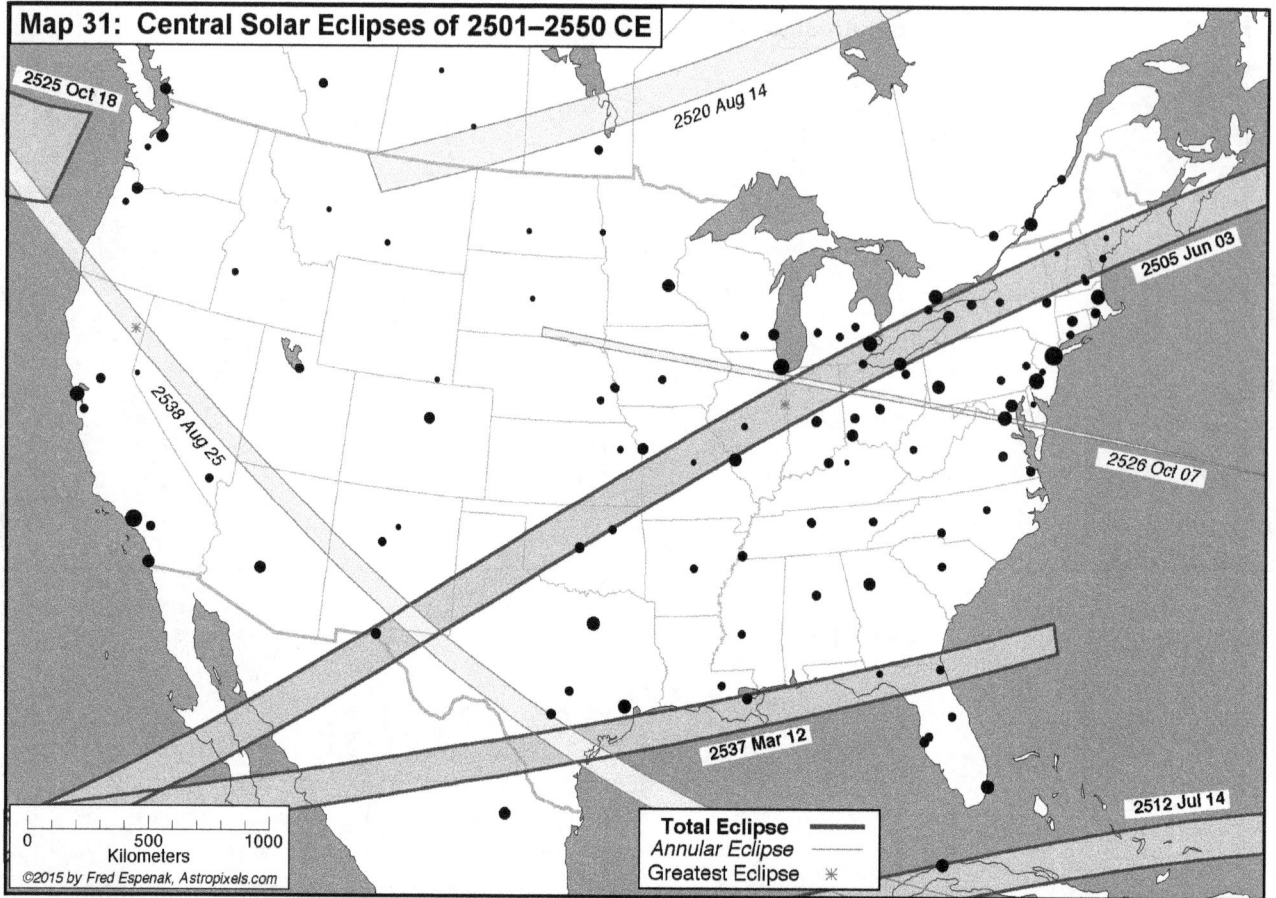

Map 31: Central Solar Eclipses of 2501–2550 CE

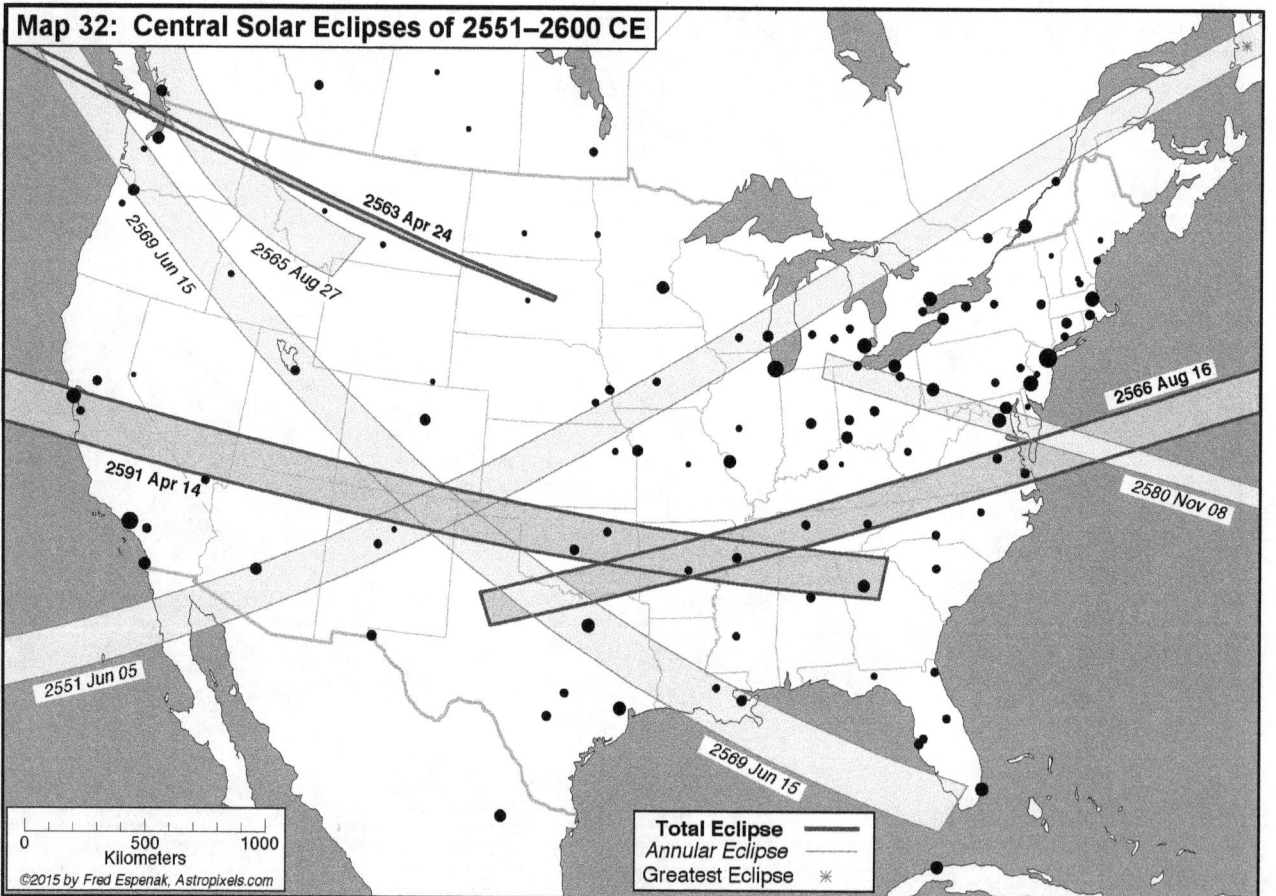

Map 32: Central Solar Eclipses of 2551–2600 CE

Map 33: Central Solar Eclipses of 2601–2650 CE

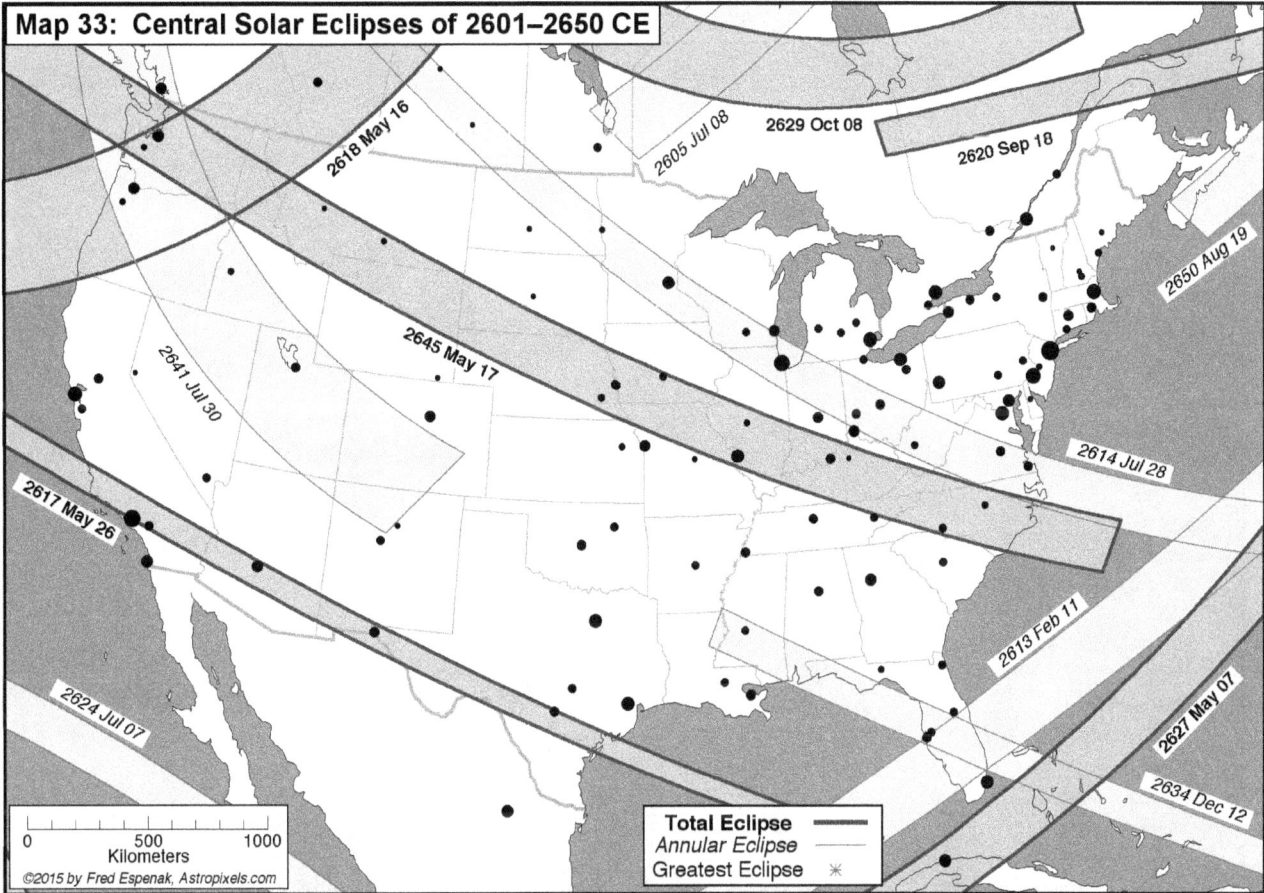

2618 May 16

2605 Jul 08

2629 Oct 08

2620 Sep 18

2650 Aug 19

2641 Jul 30

2645 May 17

2614 Jul 28

2617 May 26

2613 Feb 11

2627 May 07

2624 Jul 07

2634 Dec 12

Total Eclipse	———
Annular Eclipse	———
Greatest Eclipse	✳

0   500   1000
Kilometers
©2015 by Fred Espenak, Astropixels.com

Map 34: Central Solar Eclipses of 2651–2700 CE

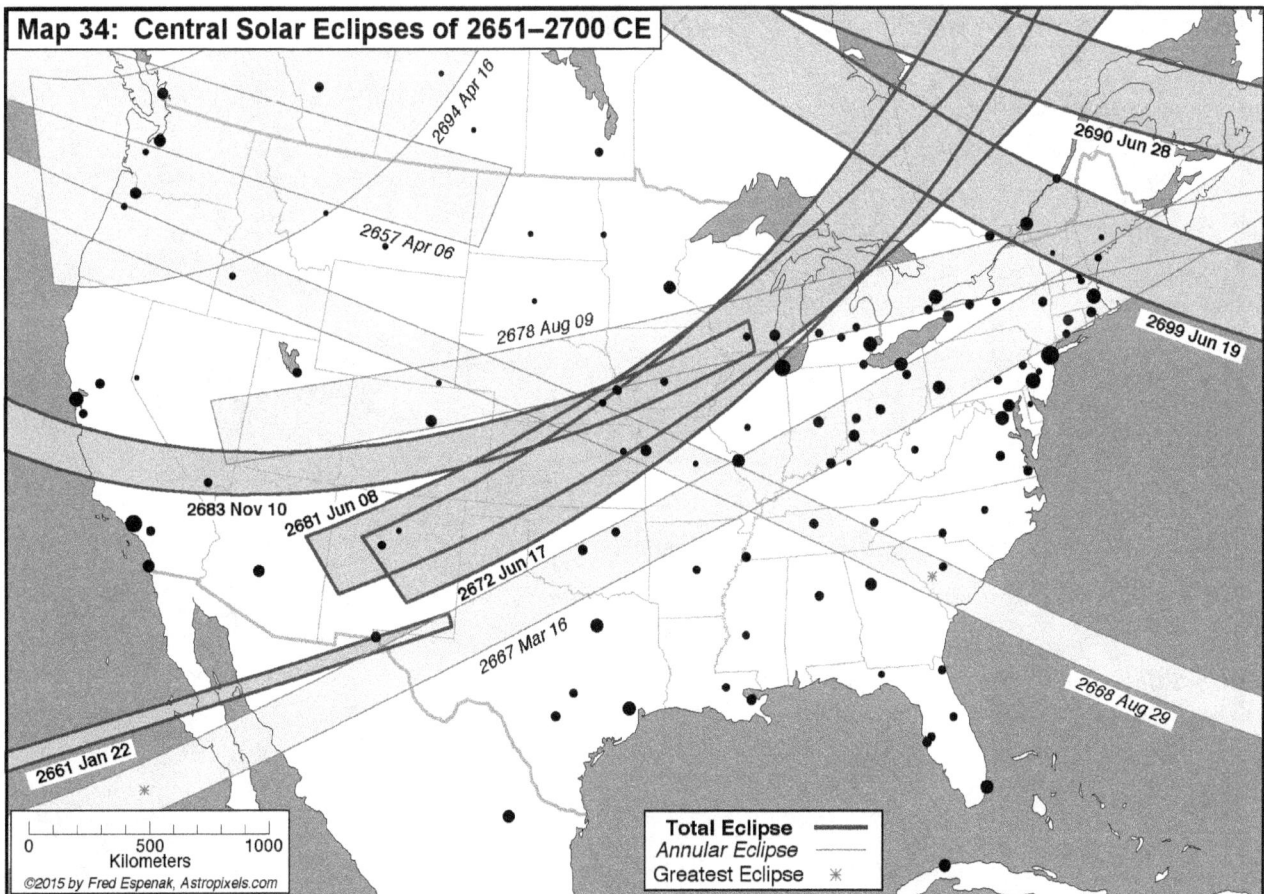

2694 Apr 16

2690 Jun 28

2657 Apr 06

2678 Aug 09

2699 Jun 19

2683 Nov 10

2681 Jun 06

2672 Jun 17

2667 Mar 16

2668 Aug 29

2661 Jan 22

Total Eclipse	———
Annular Eclipse	———
Greatest Eclipse	✳

0   500   1000
Kilometers
©2015 by Fred Espenak, Astropixels.com

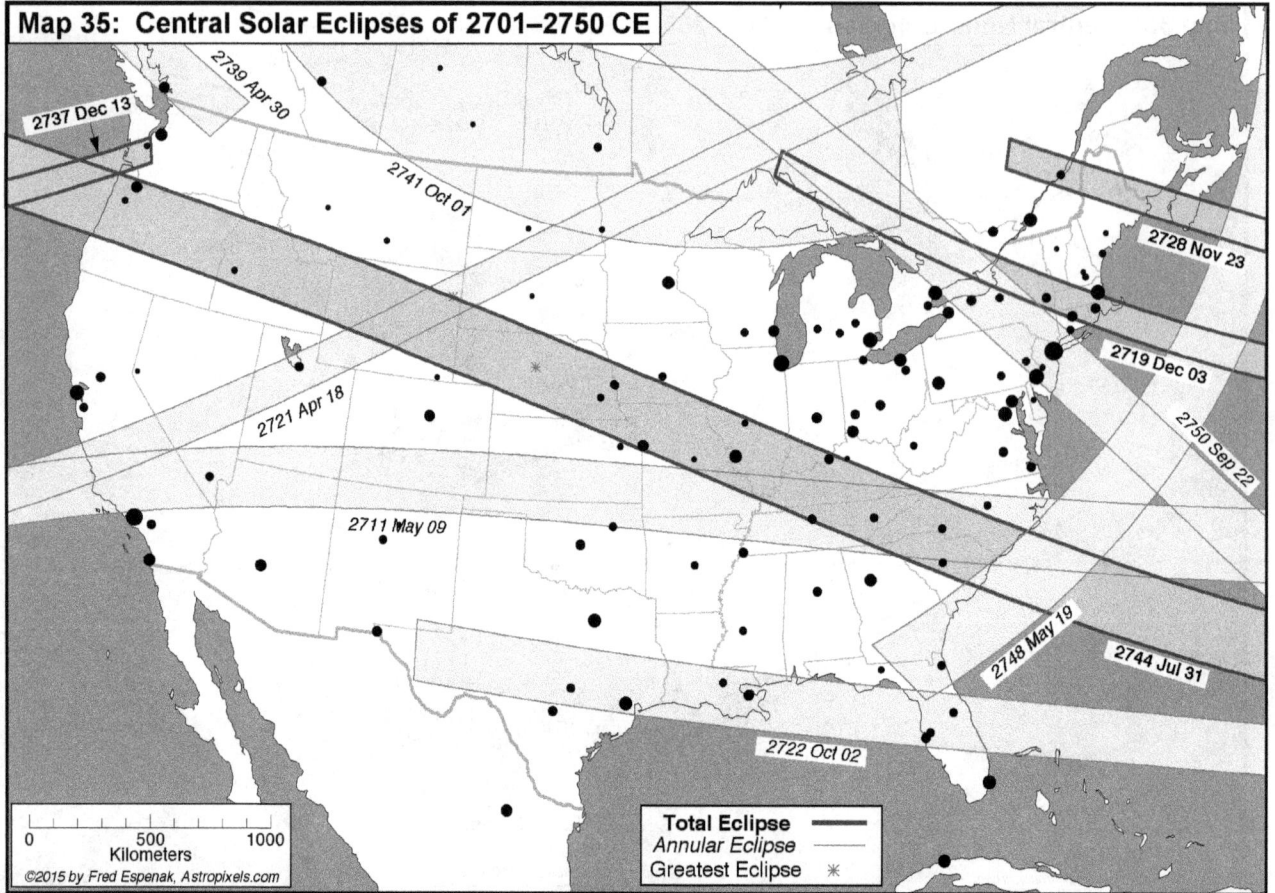

Map 35: Central Solar Eclipses of 2701–2750 CE

2739 Apr 30
2737 Dec 13
2741 Oct 01
2728 Nov 23
2719 Dec 03
2750 Sep 22
2721 Apr 18
2711 May 09
2748 May 19
2744 Jul 31
2722 Oct 02

0	500 1000
Kilometers	

©2015 by Fred Espenak, Astropixels.com

Total Eclipse ———
Annular Eclipse ———
Greatest Eclipse ✳

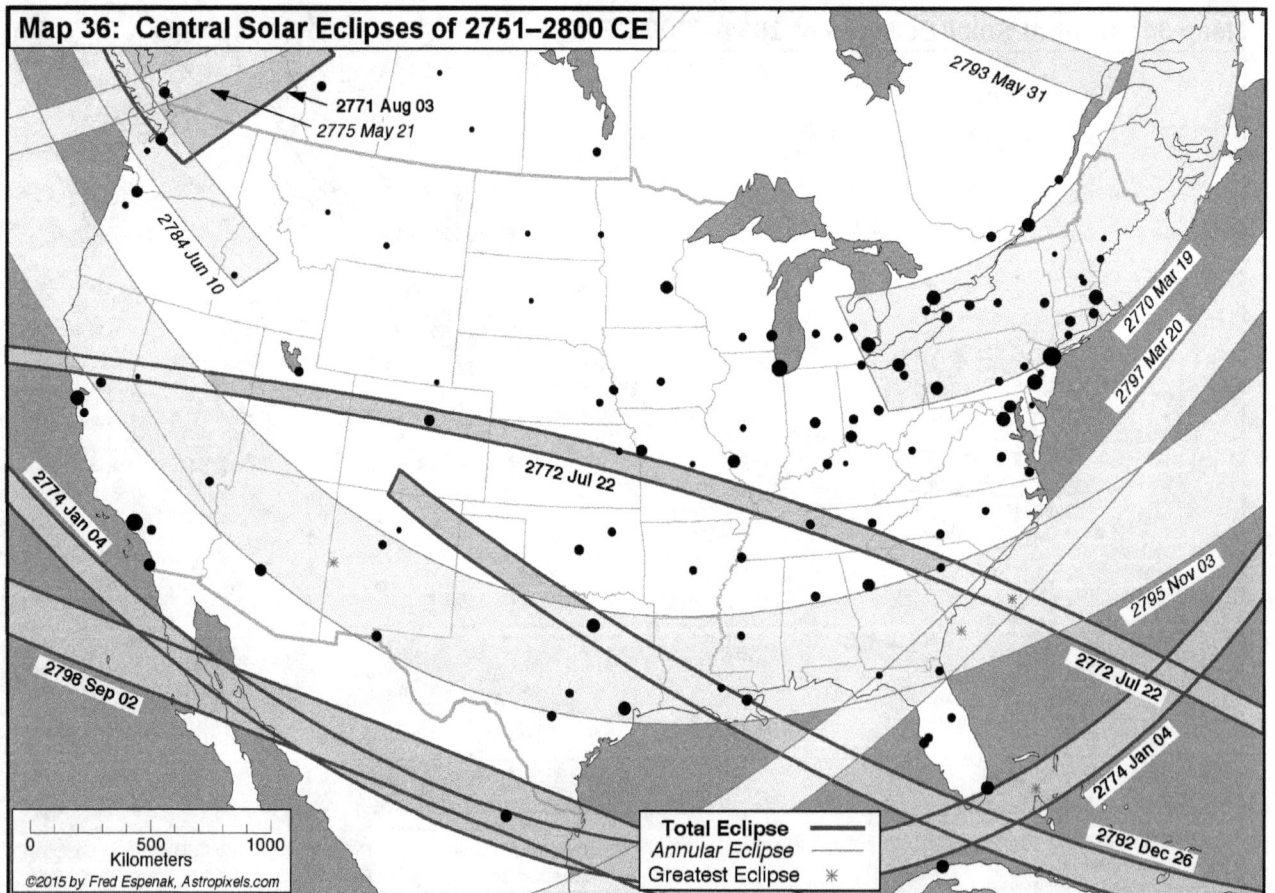

Map 36: Central Solar Eclipses of 2751–2800 CE

2793 May 31
2771 Aug 03
2775 May 21
2784 Jun 10
2770 Mar 19
2797 Mar 20
2772 Jul 22
2774 Jan 04
2795 Nov 03
2772 Jul 22
2798 Sep 02
2774 Jan 04
2782 Dec 26

0	500 1000
Kilometers	

©2015 by Fred Espenak, Astropixels.com

Total Eclipse ———
Annular Eclipse ———
Greatest Eclipse ✳

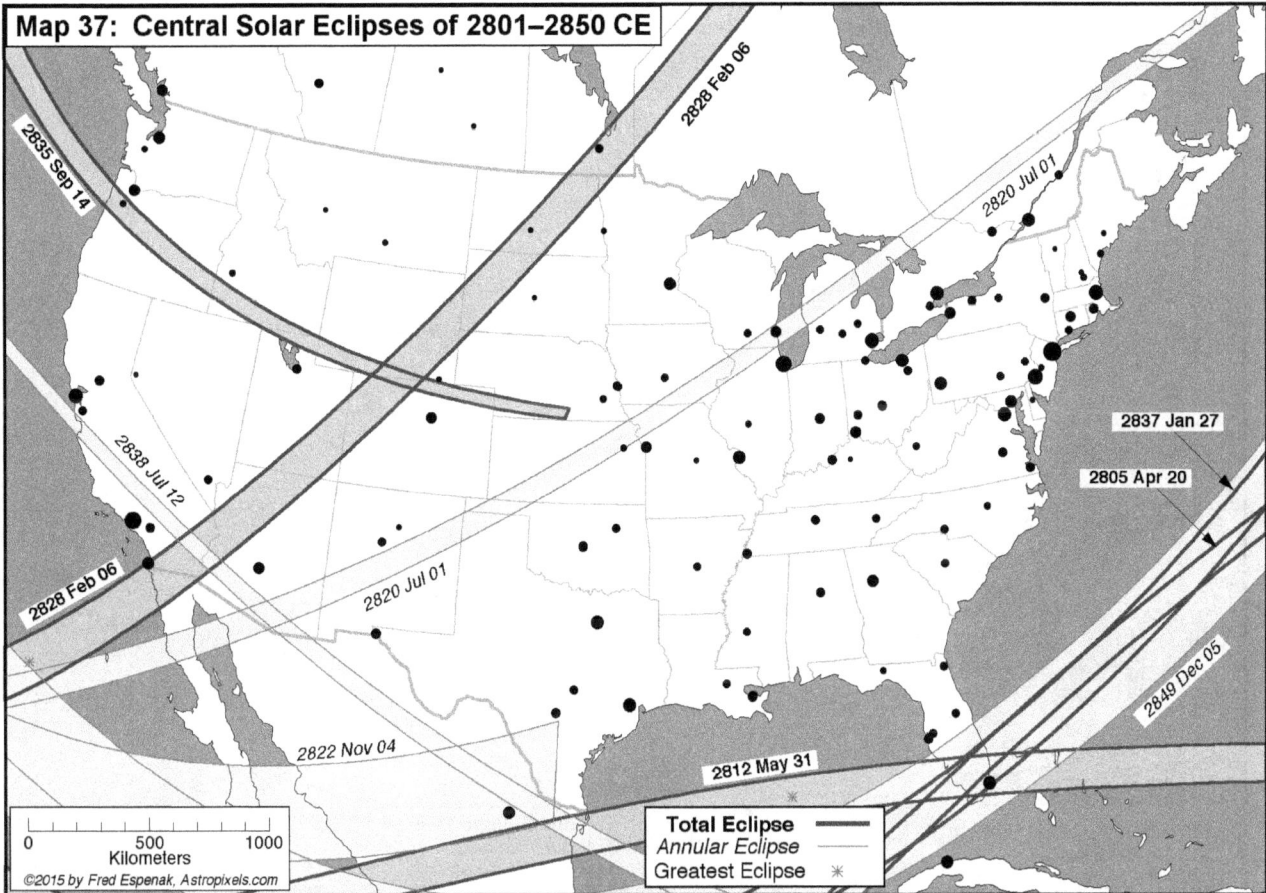

Map 37: Central Solar Eclipses of 2801–2850 CE

2828 Feb 06

2835 Sep 14

2820 Jul 01

2838 Jul 12

2837 Jan 27

2805 Apr 20

2828 Feb 06

2820 Jul 01

2849 Dec 05

2822 Nov 04

2812 May 31

Total Eclipse	——	
Annular Eclipse		
Greatest Eclipse	✳	

0   500   1000
Kilometers
©2015 by Fred Espenak, Astropixels.com

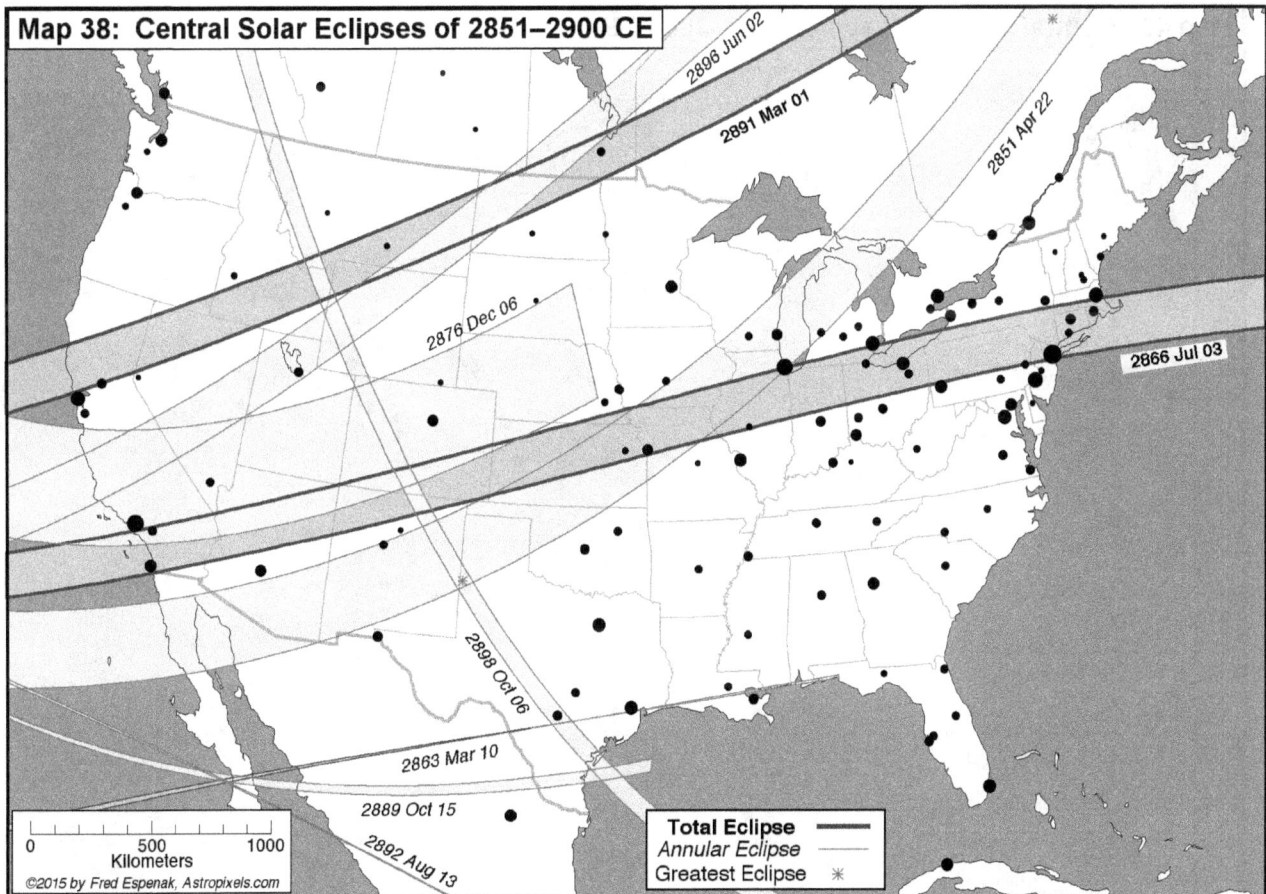

Map 38: Central Solar Eclipses of 2851–2900 CE

2896 Jun 02

2891 Mar 01

2851 Apr 22

2876 Dec 06

2866 Jul 03

2898 Oct 06

2863 Mar 10

2889 Oct 15

2892 Aug 13

Total Eclipse	——	
Annular Eclipse		
Greatest Eclipse	✳	

0   500   1000
Kilometers
©2015 by Fred Espenak, Astropixels.com

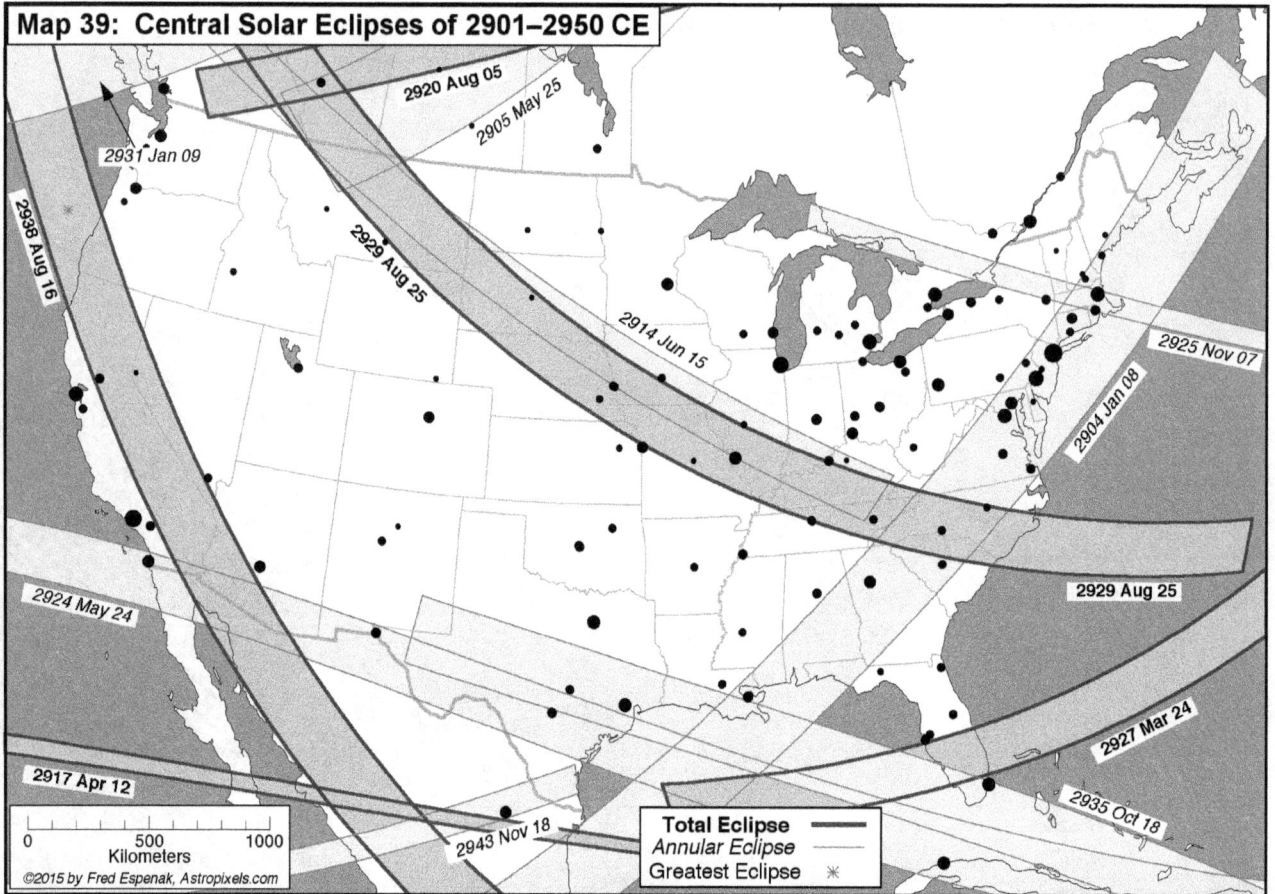

Map 39: Central Solar Eclipses of 2901–2950 CE

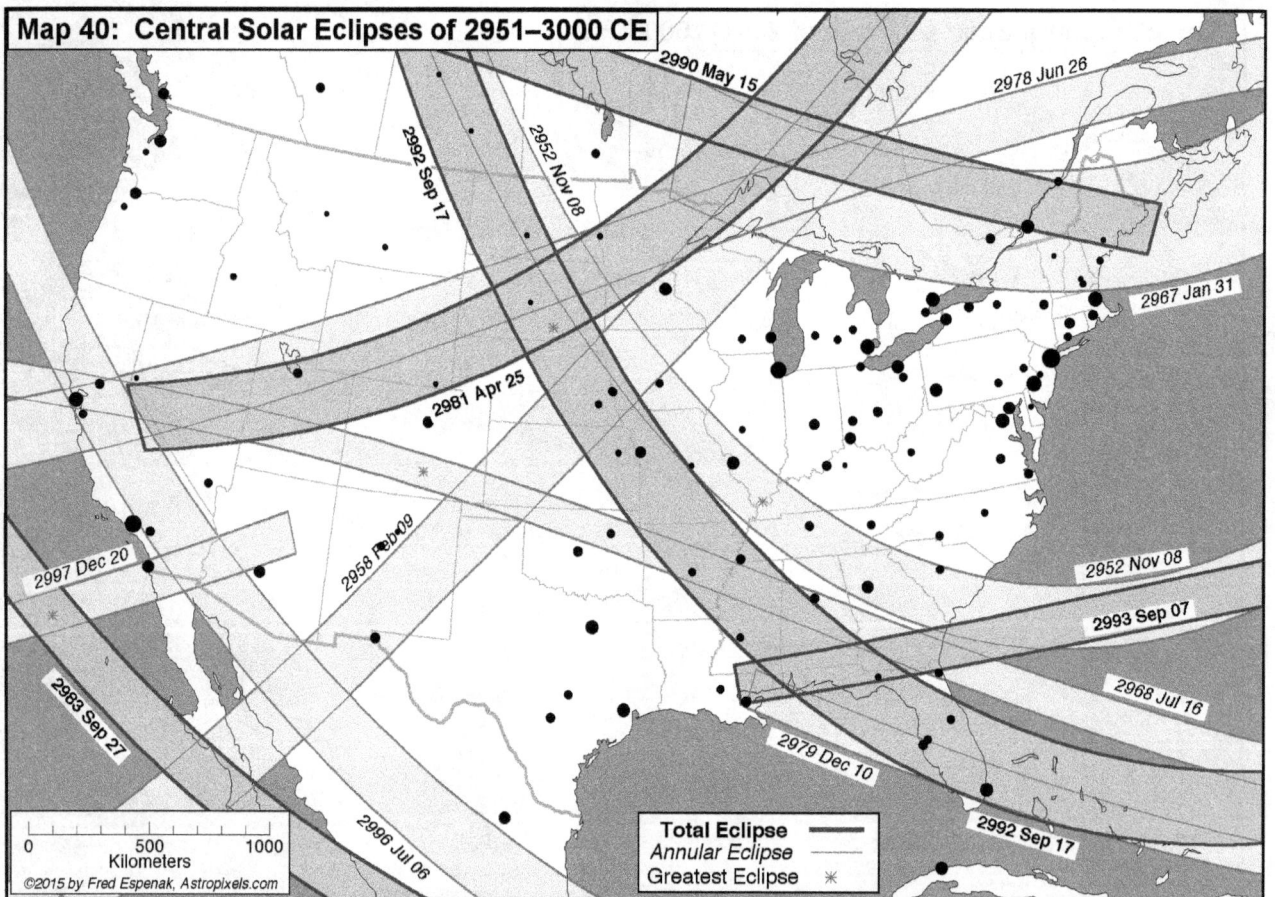

Map 40: Central Solar Eclipses of 2951–3000 CE

Bibliography

Astronomical Almanac for 2011, Washington: US Government Printing Office; London: HM Stationery Office (2010).

Espenak, F., and Meeus, J., *Five Millennium Canon of Solar Eclipses: –1999 to +3000 (2000 BCE to 3000 CE)*, NASA Tech. Pub. 2006–214141, NASA Goddard Space Flight Center, Greenbelt, Maryland (2006).

Espenak, F., and Meeus, J., *Five Millennium Catalog of Solar Eclipses: –1999 to +3000 (2000 BCE to 3000 CE)*, NASA Tech. Pub. 2006–214174, NASA Goddard Space Flight Center, Greenbelt, Maryland (2009c).

Espenak, F., *Thousand Year Canon of Solar Eclipses: 1501 to 2500*, Astropixels Pub., Portal, Arizona, 294 pp (2014).

Espenak, F., and Anderson, J., *Eclipse Bulletin: Total Solar Eclipse of 2017 August 21*, Astropixels Pub., Portal, Arizona, 157 pp (2015).

Espenak, F., *Road Atlas for the Total Solar Eclipse of 2017*, Astropixels Pub., Portal, Arizona, 50 pp (2015).

Explanatory Supplement to the Ephemeris, H.M. Almanac Office, London (1974).

Littmann, M., Espenak, F., and Willcox, K., *Totality—Eclipses of the Sun*, 3rd Ed., Oxford University Press, New York (2008).

Meeus, J., Grosjean, C.C., and Vanderleen, W., *Canon of Solar Eclipses*, Pergamon Press, Oxford, United Kingdom (1966).

Meeus, J., *J. British Astron. Assoc.,* Vol. 92, pp. 124-126 (1982).

Morrison, L., and Stephenson, F.R., "Historical Values of the Earth's Clock Error DT and the Calculation of Eclipses," *J. Hist. Astron.,* Vol. 35 Part 3, August 2004, No. 120, pp, 327–336 (2004).

Stephenson, F.R., *Historical Eclipses and Earth's Rotation*, Cambridge University Press, Cambridge (1997).

van den Bergh, *Periodicity and Variation of Solar (and Lunar) Eclipses*, Tjeenk Willink, and Haarlem, Netherlands (1955).

Astropixels Publications

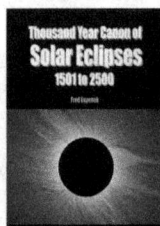

Thousand Year Canon of Solar Eclipses 1501 to 2500 (Fred Espenak) contains maps and data for each of the 2,389 solar eclipses occurring over the ten-century period centered on the present era. A comprehensive catalog lists the essential characteristics of each eclipse while a series a global maps show the exact geographic extent of each eclipse.

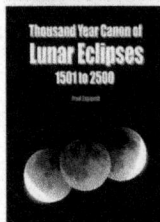

The Thousand Year Canon of Lunar Eclipses 1501 to 2500 (Fred Espenak) contains diagrams, maps and data for each of the 2,424 lunar eclipses occurring over the ten-century period centered on the present era. A comprehensive catalog lists the essential characteristics of each eclipse while a series a diagrams and maps illustrate the Moon-shadow geometry and geographic regions of visibility of each eclipse.

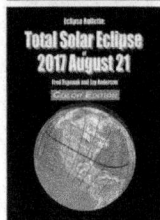

Eclipse Bulletin: Total Solar Eclipse of 2017 August 21 (Fred Espenak & Jay Anderson) is the ultimate guide to this highly anticipated event. The bulletin is a treasure trove of facts on every conceivable aspect of the eclipse. The exact details about the path of totality can be found in a series of tables containing geographic coordinates, times, altitudes, and more. Detailed maps plot the total eclipse path across the USA. Local circumstances tables for over 1000 cities provide times of each phase of the eclipse along with the eclipse magnitude, duration and Sun's altitude.

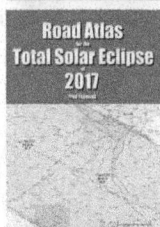

Road Atlas for the Total Solar Eclipse of 2017 (Fred Espenak) contains a comprehensive series of 37 high resolution maps of the path of totality across the USA. The large scale (1:700,000 or 1 inch = 11 miles) shows both major and minor roads, towns and cities, rivers, lakes, parks, national forests, wilderness areas and mountain ranges. The duration of totality is plotted in 20-second steps, making it easy to estimate the length of the total eclipse from any location in the eclipse path.

TOTAL Eclipse or Bust! A Family Road Trip (Patricia Espenak & Fred Espenak) is a book for the entire family. The story follows a typical family on a road trip to see the 2017 total eclipse of the Sun. Along the way the children learn all about the how and why of eclipses in a friendly and an uncomplicated way. The book also provides basic information about how to view a total solar eclipse and where to go for America's great eclipse on August 21, 2017.

The last three publications have been recognized by *Sky & Telescope magazine* as **Hot Products for 2016**. For complete details and ordering information on the above ***Astropixels Publications***, visit:

astropixels.com/pubs/

www.ingramcontent.com/pod-product-compliance
Lightning Source LLC
Chambersburg PA
CBHW051226200326
41519CB00025B/7265